Synthesis Lectures on Mathematics & Statistics

Series Editor

Steven G. Krantz, Department of Mathematics, Washington University, Saint Louis, MO, USA

This series includes titles in applied mathematics and statistics for cross-disciplinary STEM professionals, educators, researchers, and students. The series focuses on new and traditional techniques to develop mathematical knowledge and skills, an understanding of core mathematical reasoning, and the ability to utilize data in specific applications.

Subhash C. Basak

Editor

Mathematical Descriptors of Molecules and Biomolecules

Applications in Chemistry, Drug Design, Chemical Toxicology, and Computational Biology

Editor
Subhash C. Basak
University of Minnesota
Duluth, MN, USA

ISSN 1938-1743   ISSN 1938-1751   (electronic)
Synthesis Lectures on Mathematics & Statistics
ISBN 978-3-031-67840-0   ISBN 978-3-031-67841-7   (eBook)
https://doi.org/10.1007/978-3-031-67841-7

This Springer imprint is published by the registered company Springer Nature Switzerland AG
The registered company address is: Gewerbestrasse 11, 6330 Cham, Switzerland

If disposing of this product, please recycle the paper.

Preface

*Only those who will risk going too far can possibly find out how far
one can go.*

—*T. S. Eliot*

*...shall we stay our upward course? In that blessed region of Four
Dimensions, shall we linger at the threshold of the Fifth, and not
enter therein? Ah, no! Let us rather resolve that our ambition shall
soar with our corporal ascent. Then, yielding to our intellectual
onset, the gates of the Six Dimension shall fly open; after that a
Seventh, and then an Eighth...*

—*Edwin Abbott, In: Flatland*

*In science there is and will remain a Platonic element which could
not be taken away without ruining it. **Among the infinite diversity
of singular phenomena science can only look for invariants**.*

—*Jacques Monod*

*Upon this gifted age, in its dark hour,
Rains from the sky a meteoric shower
Of facts... they lie unquestioned, uncombined.
Wisdom enough to leech us of our ill
Is daily spun; but there exists no loom
To weave it into fabric...*

—*Edna St. Vincent Millay*

We are now living in an age when numerous spheres of science, technology, and life are
experiencing an explosion of huge amounts of data. The buzzword is *"data is the new oil."*
The same is true in the advancing frontiers of chemoinformatics and bioinformatics. More
and more physicochemical, biological, omics (genomics, proteomics, etc.), and sequence

data are coming to the public domain almost daily. Just as the refinement of crude oil is necessary for our daily use, the raw data in chemoinformatics and bioinformatics needs to be systematized and used in the formulation of robust predictive models to aid basic research and to assist decision support systems in socially and economically important fields like new drug discovery, and regulation of chemicals for the protection of human health and ecology.

In many cases, experimental data come with associated information about the structure of molecules or biomolecules. Unless descriptors used in model development are all totally experimentally determined, nonempirical descriptors calculated from structure without the input of any other experimental data are used for modeling. Such descriptors are mainly topological, three-dimensional (3-D), or quantum chemical in nature. Mathematical methods like matrix theory, graph theory, and information theory are some frequently used tools in molecular descriptor calculation. *The use of descriptors in predictive models arises from the dictum: Function (property) follows form (structure).* These days, often the number of predictors/descriptors (p) is much larger than n, the number of data points to be modeled. In such rank-deficient cases, proper statistical and machine learning methods need to be used to develop robust predictive models. *Therefore, it is evident that mathematical and statistical/ML methods constitute an indispensable link between structural/property data on the one hand and implementable models in the end user's computer, on the other.*

In the first chapter "Approaching Modeling in Chemoinformatics and Bioinformatics Using Mathematical Descriptors: Some Comments on the Emerging Landscape and Future Directions" of the book, Subhash C. Basak discusses the fundamental philosophy behind the formulation of chemodescriptors and biodescriptors. He also explains the methodologies for the creation of robust predictive models under rank-deficient situations. In the second chapter "Hierarchy of Descriptors: From Topology to Bio-descriptors", Marjan Vračko and Subhash C. Basak discuss the hierarchy of molecular descriptors derived from diverse mathematical techniques. The third chapter "Chirality Descriptors for Numerical Characterization of Enantiomers and Diastereomers", authored by Ramanathan Natarajan, Subhash C. Basak, and Claudiu N. Lungu, summarizes the recent development of a set of novel chirality descriptors to mathematically characterize relative chirality of molecules containing one or more asymmetric centers. The fourth chapter "QSAR Modeling Using Molecular Fragment Descriptors" by Suman K. Chakravarti delves into the use of structural fragments in the formulation of models useful in predictive pharmacology and toxicology. The fifth chapter "Quantitative Structure-Activity Analysis Using Conceptual DFT and Information Theory-based Descriptors" of the book, authored by Arpita Poddar, Ranita Pal, Shanti Gopal Patra, and Pratim Kumar Chattaraj, discusses the calculation and utility of quantum chemical density functional theory (DFT) descriptors in Quantitative Structure-Activity Analysis (QSAR) model development. The sixth chapter of the book entitled "Complexity of Molecular Ensembles with Basak's Indices: Applying Structural Information Content", jointly authored by Denis Sabirov, Alexandra Zimina, and Igor Shepelevich, discusses the utility of the novel information theoretic topological index, structural information content (SIC), originally developed by Subhash C. Basak, in the characterization of the complexity of molecular ensembles. Chapter seven entitled "Descriptors from Calculated Stereo-Electronic Properties and Molecular Electrostatic Potentials (MEPs) May Provide a Powerful "Interaction Pharmacophore" for Drug Discovery", by Apurba K. Bhattacharjee, discusses calculation and practical use of computed interaction pharmacophore in the design of new life-saving drugs. The eighth and final chapter of the book "Network-Based Molecular Descriptors for Protein Dynamics and Allosteric Regulation", by Ziyun Zhou, Lorenza Pacini, Laurent Vuillon, Claire Lesieur, and Guang Hu, deals with the use of network theory in the characterization of allosteric properties of proteins from their sequence and structural data.

I would like to specially mention that currently the fields of chemoinformatics and bioinformatics are witnessing a huge explosion of data describable by four V's: volume, velocity, variety, and veracity. However, the data per se is of very limited use unless we complete the three-step process: data is first converted to information which is then transformed into useful knowledge.

The real challenge lies in the implementation of the last two steps of the three-step pathway in the creation of actionable knowledge.

The chapters in this book involving chemodescriptors and biodescriptors expand in two distinct directions. On the one hand, the authors delve into more subtle and novel aspects of the structure of molecules and biomolecules using innovative methods ranging from topology and graph theory, and network theory to high-level quantum chemistry. Such research may lead to a new understanding of behaviors of molecules and biomolecules at the molecular and submolecular levels. On the other hand, following the "*diversity begets diversity paradigm*", in the day-to-day practical applications by end users, the collection of descriptors taken from diverse sources may be useful in the evaluation of chemicals and biochemicals, real or hypothetical, facilitated by computer algorithms and statistical/ machine learning-based software.

We sincerely hope that this book will enrich the expanding frontiers of chemoinformatics and bioinformatics in the areas of both basic and applied research.

Duluth, Minnesota, USA   Subhash C. Basak

Contents

Approaching Modeling in Chemoinformatics and Bioinformatics Using Mathematical Descriptors: Some Comments on the Emerging Landscape and Future Directions

Subhash C. Basak

Abstract

This chapter briefly discusses the history, status, and use of different classes of chemodescriptors and biodescriptors, both experimentally determined and computed from structure, in the formulation of predictive models. Possible future uses of the various classes of descriptors in model building at this age of big data analytics are highlighted.

Keywords

Molecular structure • Model object • Theoretical model • Chemodescripors • Biodescriptor • Quantitative structure–activity relationship (QSAR) • Statistical and machine learning methods • Big data analytics • Graph invariant • Topological index

1 Introduction

The most beautiful thing we can experience is the mysterious. It is the source of all true art and all science. He to whom this emotion is a stranger, who can no longer pause to wonder and stand rapt in awe, is as good as dead: his eyes are closed.

— Albert Einstein

It's all a series of serendipities

with no beginnings and no ends.

S. C. Basak (✉)
Department of Chemistry and Biochemistry, Retired Adjunct Professor, University of Minnesota Duluth, Duluth, MN 55811, USA
e-mail: sbasak@d.umn.edu

Such infinitesimal possibilities

Through which love transcends.

— Ana Claudia Antunes, The Tao of Physical and Spiritual

[T]he prepared mind requires unfettered opportunity to recognize and follow unplanned paths . . . when we pursue our passion to master what was once unknowable, we move from a plodding struggle with nature to an ongoing, enlightening conversation.

Joshua Lederberg, 21stC: Research at Columbia, 1995

Computed mathematical descriptors of chemical and biological systems, called chemodescriptors and biodescriptors [1–4, 6, 7, 16, 17, 21, 23, 31] play a key role in many basic and applied aspects of chemical and biological research related to new drug discovery and computational toxicology. In the realms of pharmaceutical drug design and assessment of toxicity of environmental pollutants as well as industrial chemicals we must screen many thousands of candidate substances for their potential beneficial properties, adverse effects, and bioactivities [3, 8]. The exhaustive testing of all these large number of chemicals in the laboratory is prohibitively costly and will necessitate the sacrifice of a huge number of test organisms. Under such circumstances, property/ bioactivity prediction models based on existing test data can act as a decision support system in the allocation of scant resources in the necessary lab testing phase [3, 6, 7, 16]. During the last half century or so, numerous software for the calculation of molecular descriptors (Table 1) have become available to us, e.g., POLLY [11, 12], Triplet [10], MolConnZ [39], Dragon [25], MOPAC [40], Gaussian [27]. Fortunately, during the past few decades a lot of property/ bioassay databases have come to the public domain (Table 2). Such data can be used for model building and prediction of properties of untested and even unsynthesized chemicals. The available computing power has been steadily increasing following Moore's law [32].

As highlighted by Basak [3], the four important pillars of quantitative structure–activity relationship (QSAR) studies supporting predictive pharmacology and computational toxicology are:

(a) Good quality experimental physicochemical and biological test data,
(b) A reasonable number of experimental data prerequisite to good model building and validation pertinent to extrapolation of models to a structurally diverse set of chemical structures [13],
(c) Availability of software that can compute properties of chemicals *from their structure only without the input of any other experimental data* so that such properties can be calculated for any molecule, real or hypothetical (Table 1),
(d) Robust statistical and machine learning methods necessary for model building and validation.

Table 1 A list of important classes of chemodescriptors and biodescriptors

	Chemodescriptors					Biodescriptors			
Physical property (experimental or calculated)	LFER/ LSER descriptors	3-D descriptors	Pharmacophoric descriptors	Graph theoretic descriptors	Quantum chemical descriptors	GC content	Proteomics descriptors	AFSDs of of DNA/RNA graphs	Descriptors of amino acid/ protein network
Boiling point/ vapor pressure; log P (octanol–water); aqueous solubility;	Hammet sigma; logP; hydrogen bond donor/ acceptor property	Molar volume; van der Waals's volume	H-bonding, hydrophobic, and electrostatic interaction sites, defined by atoms, ring centers, etc.	Different types of topostructural and topochemical indices; substructures	QC indices calculate by various levels of theory	G + C content of a genome	Mathematical descriptors d4rived from 2-D gel electrophoresis and GC/MS data	Quantitative descriptors from graphs of DNA/ RNA sequences	Protein structure descriptors from primary to the higher levels of organization

Table summary

Chemodescriptors: Linear free energy related (LFER) descriptors are derived from physical organic chemistry model systems and include hydrophobicity (e.g. Log P, octanol–water), electronic descriptors li Hammett's sigma and steric descriptors like Es. Linear solvation energy related (LSER) descriptors mainly consist of volume, polarizability, hydrogen bond donor acidity and hydrogen bond acceptor basicity. 3-D or geometrical descriptors include van der Waals' volume, surface area etc. Pharmacophore descriptors are derived from the characterization of stereo-electronic factors needed for the specific recognition of ligands by specific enzymes or recepto5rs. Graph theoretical descriptors can be either real numbers like the topological indices or substructures derived from the molecular graph model of chemical structure. Quantum chemical indices of molecules are derived by different semiempirical and ab initio methods

Biodescriptors: GC (Guanine-Cytosine) content or percentage of GC in the total base content [(G + C)/(A + T + G + C)]. Quantitative proteomics descriptors which are computed mathematically from proteomics patterns derived from 2-dimensional gel electrophoresis (2DE) or GC/ MS analysis of cellular/ tissue proteins. Alignment-free sequence descriptors (AFSDs) are mathematical descriptors derived from operations on graphs corresponding to the DNA/RNA sequences and are not related to any alignment-based approach. Quantitative descriptors of amino acid sequences of protein and 3-D structure can be used in predicting their properties

Table 2 Some available chemical and biological databases

Name	Web address	Type of information
PubChem	https://pubchem.ncbi.nlm.nih.gov	Contains data on a large number of bioassays, providing bioactivity and toxicity endpoints of chemicals
ChEMBL	ChEMBL: https://www.ebi.ac.uk/chembl/	A database with a focus on drug discovery
ToxCast	https://www.epa.gov/chemical-research/exploring-toxcast-data	A database of high-throughput screening chemical toxicity data by the US environmental protection agency
Tox 21- Toxilology in the Twenty First Century	https://ntp.niehs.nih.gov/whatwestudy/tox21/index.html	A high-throughput screening database of toxicity data by the US national institutes of health and the US environmental protection agency
Toxicity reference database (ToxRefDB	https://catalog.data.gov/dataset/toxicity-reference-database	A database of toxicity data by the US environmental protection agency
ECHA dossiers	https://echa.europa.eu/information-on-chemicals/registered-substances	European chemicals agency's (ECHA) system for registration, evaluation, authorization, and restriction of chemicals. Including various in vivo toxicity data
OECD QSAR toolbox	https://www.oecd.org/chemicalsafety/risk-assessment/oecd-qsar-toolbox.htm	A collection of QSAR models and toxicity data, developed by the organization for economic co-operation and development (OECD)
EURL ECVAM database	https://data.jrc.ec.europa.eu/dataset/b7597ada-148d-4560-9079-ab0a5539cad3	Data on alternative methods to animal testing (DB-ALM): a database of alternative methods for toxicity testing
UC Irvine (UCI) machine learning repository	https://archive.ics.uci.edu/ml/index.php	A good collection of data of all kinds, ready for model building

(continued)

Table 2 (continued)

Name	Web address	Type of information
GenBank and WGS database	https://www.ncbi.nlm.nih.gov/genbank/statistics/	Whole genome shotgun (WGS) database contains data on chromosomes of prokaryotes or eukaryotes that are generally being sequenced by whole genome shotgun strategy
RCSB protein data bank (RCSB PDB	https://www.rcsb.org/	Contains experimentally derived and computed protein structure data
GISAID (global initiative on sharing avian influenza data)	https://gisaid.org/	Provides open access genomic data of influenza viruses and the coronavirus responsible for the COVID-19 pandemic
DRIGBAML	https://www.ncbi.nlm.nih.gov/pmc/articles/PMC2238889/	Provides detailed data with comprehensive drug target and drug action information

Today, a combination of the above-mentioned four factors, in particular, gives us the opportunity of developing powerful predictive models for practical use in various areas of basic and applied research. As more experimental test data on collections of highly diverse chemical structures are being available to us, in QSAR research there is a necessary paradigm shift, *a transition from the "Congenericity Principle" toward the "Diversity begets Diversity" principle* [13].

2   The Enormous Descriptor Landscape

It is worth mentioning that In the post-genomics era, catapulted by the completion of the Human Genome Project [29], a lot of data on the macromolecular DNA, RNA, protein sequences as well as expression of genetic information in normal and chemically/ biologically/clinically affected cells/tissues are being collected in different publicly available databases (Table 2). Such data are good starting points in the development of mathematical and computational biodescriptors for the characterization of biological systems. Table 1 gives a short list of popularly used chemodescriptors and biodescriptors.

The term chemodescriptors is used for descriptors derived for small organic molecules or more complex entities like peptides [3]. These may include both experimentally determined properties like melting point, boiling point and hydrophobicity (e.g., Octanol–water partition coefficient) of chemicals. These may also include the different classes of molecular descriptors which can be computed from the structure of chemicals, e.g., topostructural

(TS), topochemical (TC), geometrical or three-dimensional (3-D) and various types of quantum chemical (QC) indices. Biodescriptor development is aimed at the characterization of complex biological objects like DNA/ RNA/ protein sequences, proteomics maps or complex macromolecular networks in cells [2]. *A descriptor, chemo- or biodescriptor, quantifies certain aspects of the structure of the entity which it characterizes.*

The fundamental philosophy of descriptor formulation consists of two major stages: (a) Representation of the entity (chemical or biological object like the molecule or DNA sequence) under investigation into a model object and (b) Characterization of the model object using mathematical models [14, 20].

In the context of molecular science, the various concepts of molecular structure (e.g. classical valence bond representations, various chemical graph-theoretic representations, ball and spoke model of a molecule, representation of a molecule by minimum energy conformation, or symbolic representation of chemical species by Hamiltonian operators) are model objects derived through different methods of abstractions of the same chemical reality. In each instance, the equivalence class (concept or model of molecular structure) is generated by selecting certain aspects while ignoring some unique properties of those actual objects. This explains the plurality of the concept of molecular structure and their autonomous nature, the word "autonomous" being used in the sense that one concept is not logically derived from the other [14, 36, 42].

Figure 1 gives a short overview of the development of physicochemical molecular descriptors since 1868:

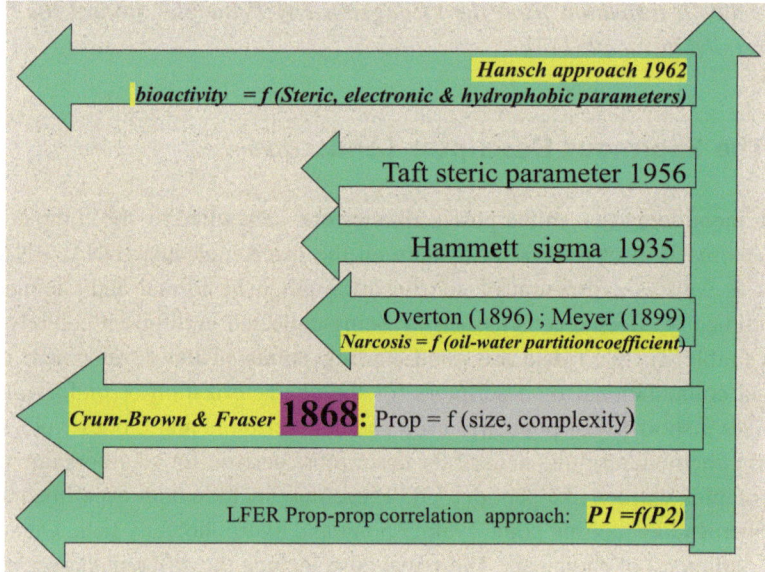

Fig. 1 Brief history of Hansch approach: 1868–to date

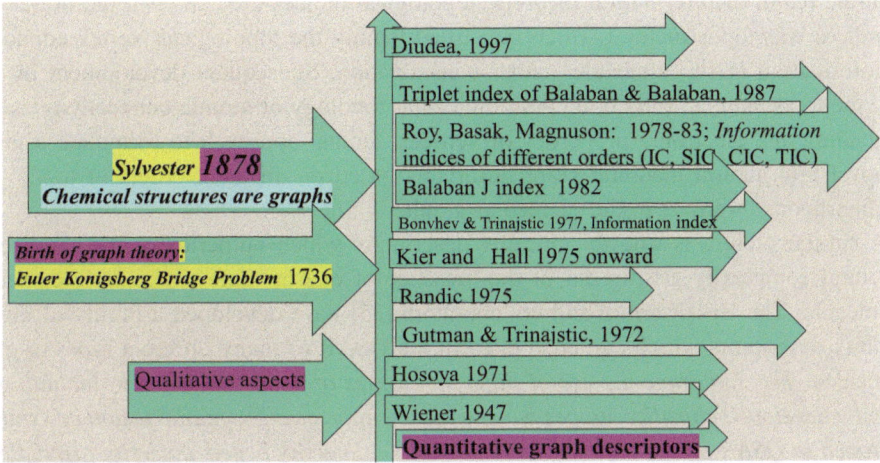

Fig. 2 A brief history of graph theory based indices: 1736–to date

Legend to Fig. 1: Fig. 1 gives a brief chronological order of the development of LFER (linear free energy related) molecular descriptors. For more details, please see Basak [6].

Figure 2 gives a bird's eye view of the history of the development of graph theory-based descriptors starting with the initial discovery by Euler in 1736.

Legend to Fig. 2: Fig. 2 gives a brief overview of the development of graph theory-based descriptors starting with the discovery of graph theory by Euler in 1736.

3   Use of Descriptors in Model Building

In recent years graph theoretical descriptors, both numerical graph invariants and substructural descriptors, have found wide applications in the prediction of physical property, pharmacological properties and potential toxicity of chemicals using QSAR models [6, 21, 31]. They have also been used in QMSA (quantitative molecular similarity analysis) models in the quantification of intermolecular similarity/ dissimilarity of chemicals for drug design and predictive toxicology [5]. Figure 2 above gives a brief history of the development and use of numerical graph theoretical descriptors, also known as topological indices, starting from the discovery of graph theory by Euler in 1736.

In many cases mathematical descriptors provide a numerical scale for some qualitative attributes of the structure. Basak [3, 4]. For example, the connectivity index developed by [37] was supposed to represent molecular branching. This was a topostructural (TS) class of descriptor. The inclusion of chemically relevant information into molecular graphs by [30] resulted in a wide variety of indices of the connectivity type which had a lot of success in QSAR.

In the realm of information theoretic topological indices [15, 18, 19], the initial crop of indices was topostructural which quantified mainly the topological/ vertex connectivity information of the molecules under consideration. Subsequent development by [38] used the topochemical (involving both molecular topology or atomic connectivity as well as bonding and electronic properties of vertices/ atoms) approach to formulate a family of novel information theoretic graph invariants based on different orders of topological neighborhoods of atoms in the molecular graph.

A similar picture is true for chirality descriptors which numerically characterize the structural complexity arising out of the presence of one or more asymmetric centers in a molecule. The Basak group and collaborators [35] have developed a family of diverse chirality descriptors which can be applied in the QSAR of many different types of chiral molecules. *The simultaneous use of different classes of chemodescriptors, including the recently developed chirality indices may be looked upon as a pragmatic fusion or synthetic approach in QSAR model building in line with the diversity begets diversity principle put forward by Basak group* [7, 13].

In the realms of biodescriptors-mathematical quantifiers of structural aspects of biomolecules like DNA/RNA sequences, proteomics maps, etc.- various indices have been developed using multiple approaches which found useful applications in the characterization of emerging global pathogens [9, 34, 41], prediction of toxicity of chemicals [2] to cells and tissues as well as design of new drugs and peptide vaccines for new viral diseases [33]. Becasuse in many cases of chemical-biological interactions only chemoinformatics cannot model the complex situation effectively, an integrated QSAR (I-QSAR) approach involving knowledge derived both from chemoinforamtics and bioinfoermtics domains seems to be the right technique for effective model formulation [6, 7, 13, 17]. The pluarlistic I-QSAR approach that uses phyhsicochemical proptrries, computed structural descriptors and biodescriptors may evolve as a viable model for predicting complex emergent properites of chemical and biiological systems [2].

4   Conclusion

Nature knows no pause in progress and development and attaches her curse on all inaction.

Johann Wolfgang von Goethe

See first, think later, then test. But always see first. Otherwise, you will only see what you were expecting. Most scientists forget that.

Douglas Adams

Data matures like wine, applications like fish.

—James Governor

Descriptors may be derived through the applications of different approaches, viz. (a) laboratory experiment to determine properties like boiling point, partition coefficient (logP) or abundance of proteins in proteomics protocols; (b) derivation of LFER descriptors from physical organic chemistry [28]. (c) quantum chemical calculations using MOPAC [40] or Gaussian (2016), (d) formulation of graph theoretic or topological descriptors of molecules [3, 16, 22, 23, 31]; (e) quantification of proteomics maps or patterns using mathematical methods [2], and characterization of DNA/RNA sequences [34].

A review of modeling literature would testify that each of the above six classes of descriptors mentioned above have often been used for model building in its respective area in a domain specific manner. New research in novel descriptor development from physicochemical or structural approaches is continuing. A new descriptor encodes novel information about a molecule or biomolecule unless it is perfectly correlated with any other preexisting molecular descriptor. *But how to best use the expanding collection of descriptors?* Whereas formulation of a new descriptor may be looked upon as *an augmentation of the descriptor space*, extraction of useful, if possible orthogonal, information from a large and diverse collection of chemo- and biodescriptors is very important.

When a set of large, say one thousand, descriptors are calculated for a small set, e.g., one hundred molecules, data for this investigation can be viewed as n = 100 vectors (chemicals) in p = 1,000 dimensions (descriptors). Each chemical is a point in R^{1000}. However, in most practical situations, since many molecular descriptors are highly correlated with one another, the 100 points in R^{1000} will lie nearly on a subspace of much lower dimension. Statistical methods like principal component analysis (PCA) or various machine learning methods can be used for the reduction of dimensionality in such cases [12, 13]. While novel mathematical methods will continue developing novel descriptors, existing and emerging statistical as well as machine learning tools will play a pivotal role in the formulation of robust predictive models. It is the understanding of this author that in the future, particularly for modeling properties and bioactivities of large and diverse sets of chemicals, collections of descriptors taken from the various classes discussed above will be used together seamlessly.

Acknowledgements The author is thankful to Gregory D. Grunwald for technical support.

References

1. Basak SC (1987) Use of molecular complexity indices in predictive pharmacology and toxicology: a QSAR approach. Med Sci Res 15:605–609
2. Basak SC (2010) Role of Mathematical chemodescriptors and proteomics-based biodescriptors in drug discovery. Drug Dev Res 72:1–9

3. Basak SC (2013) Mathematical descriptors for the prediction of property, bioactivity, and toxicity of chemicals from their structure: a chemical-cum-biochemical approach. Curr Comput-Aided Drug Des 9:449–462
4. Basak SC (2013) Philososophy of mathematical chemistry: a personal perspective. HYLE 19:3–17
5. Basak SC (2014). Molecular similarity and hazard assessment of chemicals: a comparative study of arbitrary and tailored similarity spaces. J Eng Sci Manage Educ 7(III): 178–184
6. Basak SC (2021) My tortuous pathway through mathematical chemistry and QSAR research with memories of some personal interactions and collaborations with Professors Milan Randic and Mircea Diudea. Croat Chem Acta 93(4):247–258
7. Basak SC (2021) Some comments on the three-pronged chemobiodescriptor approach to QSAR—a historical view of the emerging integration. Curr Comput Aided Drug Des (2022) (in press)
8. Basak SC, Balasubramanian K, Gute BD, Mills D, Gorczynska A, Roszak S (2003) Prediction of cellular toxicity of halocarbons from computed chemodescriptors: a hierarchical QSAR approach. J Chem Inf Comput Sci 43:1103–1109
9. Basak SC, Bhattacharjee AK, Nandy A (eds) (2019) Zika virus: basic biology, surveillance, vaccine design and anti-Zika drug discovery: computer-assisted strategies to combat the menace. Nova Science Pub Inc; 1st edn. New York
10. Basak SC, Grunwald GD, Balaban AT (1993) TRIPLET, Copyright of the Regents of the University of Minnesota
11. Basak SC, Harriss DK, Magnuson VR (1988) POLLY v. 2.3: Copyright of the University of Minnesota, USA
12. Basak SC, Magnuson VR, Niemi GJ, Regal RR (1988) Determining structural similarity of chemicals using graph theoretic indices. Discrete Appl Math 19:17–44
13. Basak SC, Majumdar S (2016) Exploring two QSAR paradigms-congenericity principle versus diversity begets diversity principle analyzed using computed mathematical chemodescriptors of homogeneous and diverse sets of chemical mutagens. Curr Compu- Aided Drug Des 12:1–3
14. Basak SC, Niemi GJ, Veith GD (1990) Optimal characterization of structure for prediction of properties. J Math Chem 4:185–205
15. Basak SC, Roy AB, Ghosh JJ (1979) Study of the structure-function relationship of pharmacological and toxicological agents using information theory. In: Avula XJR, Bellman R, Luke YL, Rigler AK (eds) Proceedings of the second international conference on mathematical modelling. University of Missouri-Rolla, Rolla, Missouri, USA, pp 851–856
16. Basak SC, Villaveces JL, Restrepo G (eds) (2015) Advances in mathematical chemistry and applications, vol 1 & 2. Elsevier & Bentham Science Publishers, Amsterdam & Boston
17. Basak SC, Vracko M (eds) (2022) Big data analytics in chemoinformatics and bioinformatics: with applications to computer-aided drug design, cancer biology, emerging pathogens and computational toxicology. Elsevier
18. Bonchev D, Trinajstic N (1977) Information theory, distance matrix, and molecular branching. J Chem Phys 67:4517–4533
19. Bonchev D (1983) Information theoretic indices for characterization of chemical structures. Research Studies Press, Chichester, United Kingdom
20. Bunge M (1973) Method, model and matter. Dordrecht: D. Reidel, pp. iii+196. D.Fl. 45
21. Chakravarti SK (2018) Distributed representation of chemical fragments. ACS Omega 3(3):2825–2836. https://doi.org/10.1021/acsomega.7b02045
22. Dehmer M, Basak SC (eds) (2012) Statistical and machine learning approaches for network analysis. In: Dehmer M, Basak SC (eds) Wiley, Hoboken, New Jersey, USA

23. Devillers J, Balaban AT (eds) (1999) Topological indices and related descriptors in QSAR and QSPR. Gordon and Breach, Amsterdam, The Netherlands

24. DRAGON 7.0. https://chm.kode-solutions.net/pf/dragon-7-0/. Accessed 7 Feb 2003

25. Dragon (2028) On-line information about Dragon software is given at the web site http://www.talete.mi.it/products/dragon_description.htm

26. Euler I (1736) Solutio problematis ad geometriam situs pertinentis. Comment Acad Sci U. Petrop 8: 128–140

27. Gaussian 09, Revision A.02 (2016) Frisch MJ, Trucks GW, Schlegel HB, Scuseria GE, Robb MA, Cheeseman JR, Scalmani G, Barone V, Petersson GA, Nakatsuji H, Li X, Caricato M, Marenich A, Bloino J, Janesko BG, Gomperts R, Mennucci B, Hratchian HP, Ortiz JV, Izmaylov AF, Sonnenberg JL, Williams-Young D, Ding F, Lipparini F, Egidi F, Goings J, Peng B, Petrone A, Henderson T, Ranasinghe D, Zakrzewski VG, Gao J, Rega N, Zheng G, Liang W, Hada M, Ehara M, Toyota K, Fukuda R, Hasegawa J, Ishida M, Nakajima T, Honda Y, Kitao O, Nakai H, Vreven T, Throssell K, Montgomery JA, Jr. Peralta JE, Ogliaro F, Bearpark M, Heyd JJ, Brothers E, Kudin KN, Staroverov VN, Keith T, Kobayashi R, Normand J, Raghavachari K, Rendell A, Burant JC, Iyengar SS, Tomasi J, Cossi M, Millam JM, Klene M, Adamo C, Cammi R, Ochterski JW, Martin RL, Morokuma K, Farkas O, Foresman JB, Fox DJ, Gaussian, Inc., Wallingford CT.

28. Hansch C, Leo A (1996) Exploring QSAR: fundamentals and applications in chemistry and biology. American Chemical Society, Washington, DC

29. Human Genome Project (2003). https://www.genome.gov/human-genome-project. Accessed 14 Feb 2023

30. Kier LB, Hall LH (1976) Molecular connectivity in chemistry and drug research. Academic Press, New York

31. Klopman G (1984) Artificial intelligence approach to structure-activity studies. Computer automated structure evaluation of biological activity of organic molecules. J Am Chem Soc 106(24): 7315–21. https://doi.org/10.1021/ja00336a004

32. Moore's law. https://www.investopedia.com/terms/m/mooreslaw.asp. Accessed 7 Feb 2023

33. Nandy A, Basak SC (2015) Prognosis of possible reassortments in recent H5N2 epidemic influenza in USA: implications for computer-assisted surveillance as well as drug/vaccine design. Curr Comput Aided Drug Des 11:110–116

34. Nandy A, Harle M, Basak SC (2006) Mathematical descriptors of DNA sequences: development and application. ARKIVOC 9:211–238

35. Natarajan R, Lungu CN, Basak SC (2024) Chirality descriptors for structure–activity relationship modeling of bioactive molecules. J Math Chem. https://doi.org/10.1007/s10910-023-01531-2

36. Primas H (1981) Chemistry, quantum mechanics, and reductionism. Springer, Berlin

37. Randic M (1975) Characterization of molecular branching. J Am Chem Soc 7:6609–6615

38. Roy AB, Basak SC, Harriss DK, Magnuson VR (1984) Neighborhood complexity and symmetry of chemical graphs and their biological applications. In: Avula XJR, Kalman RE, Liapis A, Rodin EY (eds) Mathematical modeling in science and technology. Pergamon Press, New York, pp 745–750

39. MolconnZ (2003) Version 4.05, Hall Ass Consult. Quincy, MA

40. MOPAC (2016) http://openmopac.net/MOPAC2016.html

41. Vracko M, Basak SC, Dey T, Nandy A (2021) Cluster analysis of coronavirus sequences using computational sequence descriptors: with applications to SARS, MERS, and SARS-CoV-2 (COVID-19). Curr Comput Aided Drug Des 17:936–945

42. Weininger SJ (1984) The Molecular structure conundrum: can classical chemistry be. reduced to quantum chemistry? J Chem Ed 61:939–944

Hierarchy of Descriptors: From Topology to Bio-descriptors

Marjan Vračko and Subhash C. Basak

Abstract

Molecular descriptors and biodescriptors quantify different structural aspects of small molecules and biomolecules. Different mathematical approaches are used to quantify salient aspects of molecular and biomolecular strucuraes. This chapter discusses the nature and methods of calculations of various classes of chemodescripotors and biodescriptors developed and used by reseaerchers during the past century.

Keywords

Quantitqative structure–activity relationship (QSAR) • AOP (adverse outcome pathway) • Topological indices • Adjacency matrix • Distance matrix • Connectivity index • Balaban J index • Information theoretic indices • Electrotopological state indices • Valence electron mobile (VEM) count • Hartree–Fock approximation • Electron density functional approximation • Highest occupied molecular orbital (HOMO) • Lowest unoccupied molecular orbital (LUMO) • Molecular fragment • Kohonen artificial neural networks (KANN) • Principal component analysis (PCA) • Genetic algorithm (GA) • Euklidian distance • Structural similarity

M. Vračko (✉)
Kemijski Inštitut/National Institute of Chemistry, Ljubljana, Slovenia
e-mail: marjan.vracko@ki.si

S. C. Basak
1802 Standord Avenue, Duluth, MN 55811, USA

1   Introduction

The basic question in computational chemistry science is how to present a chemical structure in a language 'understandable to computers'. Usually the human view of a molecule
is a structure drawn as a two-dimensional sketch. Its numerical description is one of central tasks in computational chemistry. Nowadays, thousands of numerical descriptors can
be easily calculated from structure, which can be used in quantitqative structure–activity
relationship (QSAR) models or analysed with different numerical methods. In this chapter
we present the hierarchy of molecular desriptors.

A decade ago, the AOP (adverse outcome pathway) was introduced as a new concept
in toxicology of chemicals [1, 2]. The AOP conceptually describes the sequence of events
on different levels of biological organisation from interaction between a xenobiotic and
bio-molecules, to cells, tissues, organs, and organism, which leads at the end to dis-
functioning of organisms or changing in populations. It is schematically shown in Fig. 1.
QSAR models including other numerical methods are applicable along entire AOP chain.
The ultimate goal of modern computational toxicology including QSAR modeling and
other modeling techniques is to link the data, information and knowledge from different
levels in AOP scheme and get a complex picture of (eventuel) adverse effects of chemicals
on organisms.

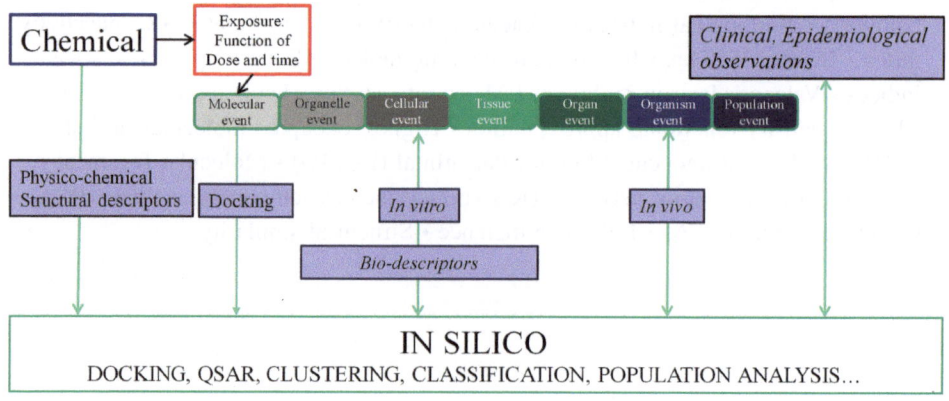

Fig. 1 The information flow from AOP scheme

2   Hierarchy of Descriptors

Molecular descriptors are parameters, which quantify certain aspects of chemical structure. There are many physico-chemical descriptors which quantify physical and chemical properties of molecules, or structural attributes which are usually calculated. Examples of physico-chemical descriptors are given in Sect. 2.1 of this chapter. The structural descriptors can be calculated solely on the basis of molecular structure. Today several thausands of descriptors can be easily calculated using available software.

2.1   One-Dimensional (1D) Descriptors

They provide the basic information about molecules, like number of atoms, molecular weight, or lipophilic properties (log P and log D). Log P is expressed as a ratio of solubility in octanol and water and it describes the readiness of a molecule to prefer the polar (water) or non-polar (octanol) environment. Similarly, the log D is expressed as a ratio between both solubilities given as function of the solutant's acidity pH. They were introduced into QSAR modeling in the pioneer works of Hansh [3] under assumption that lipofilicity determines the transport of chemicals from the site of application to the cell inside and thus basically determines the biological response of a compound. Log P (log D) can be measured or calculated using one of the many commercial or freely available computational models. Respecting the pioneer QSAR works two other empirical constants has to be mentioned: Hammet and Taft substituent constants, which describe the electronic and steric properties of molecules [3]. Other physico-chemical constants often used as descriptors are water solubility, molecule refractivity index and various partition coefficients, like partitioning coefficients between blood and brain, or other tissues, Octanol–water partition coefficient, etc.

2.2   Two-Dimensional (2D) Descriptors—Topological Indices

The 2D picture of molecular structure describes the topological property of a molecule, i.e., how the atoms are interconnected, but it does not show the metric properties, i.e., the lengths and angles between bonds. In mathematical jargon the molecules are graphs, ub wgucg the atoms are vertices and the bonds are edges. The 2D descriptors, alias 'topological indices', are deduced mainly from two matrices, first, from the adjacency matrix, which describes how the atoms are interconnected, and the topological distance matrix (D). It shows the shortest distances following the graph between all atom pairs. The first topological index was proposed by Wiener and Plat about 80 years ago. The index is a half-sum of all elements of matrix D. In the following decade the idea has been extended and many different indices have been proposed such as hyper-Wiener index,

Zagreb index, Szeged index, etc. In seventees several connectivity indices have been proposed like Randić index, Kier and Hall index (Kappa index), or Balaban index (J) [4–6]. The connectivity index considers atoms in a molecule and bonds associated to individual atoms.

An important topic in description of molecules is stereoisomerism. In 2D picture of molecular structure the stereoisomerism is usually introduced with an extra factor, which multiplies the original topological index [7–9]. For further reading about the topological indices the reference [10] is recommended.

2.3   Two-Dimensional (2D) Descriptors—Electrotopological Indices

The electrotopological indices encode, beside the topology, the electronic properties of atoms, e.g., the E-state index proposed by Kier and Hall [5]. Basicaly, the index is a weigted sum over atom contributions. They include the information about the valency of an atom and also the information about the neighboring atoms. Another approach is described in reference [11] where authors introduced extended topochemical atom (ETA) indices. The ETA scheme includes several parameters, like core count (zero for hydrogen atoms), electronegativity, and valence electron mobile (VEM) count.

2.4   Three-Dimensional (3D) Descriptors—Quantum Chemical Descriptors

The quantum mechanics plays an important role in chemistry. It is one of the methods for optimization of molecular geometry and thus for determination of 3D molecular structures. Notably, the quantum mechanics provide an insight of chemical reactions [12]. Exact quantum chemical treatment, i.e., the solving of Schrödinger equation for electrons in Coulomb field of nuclei, is not possible for the real word molecules. Indeed, the exact solution of Schrödinger equation exists only for hydrogen atom and H_2^+ ion. The problem is exact unsoluble due the nature of electron–electron interaction. In the chemistry two approximations are widely used: the Hartree–Fock approximation and the electron density functional approximation. Here, the electron–electron interaction is approximated with an average electronic potential, which is calculated on iterative way. The basic natural law, which requires the change of the wave function's sign when two electrons are interchanged, appears in the HF equation as exchange potential. Result of a calculation is orbital energies and molecular orbitals. The configuration of the electronic state is determined with the occupation of orbitals. The late ones serve as a basis for calculation of charge distribution in a molecule. The Hartree–Fock is computer intensive method and, in spite of the fast development of computers technique it still remains limited. To receive the electronic structures for a large bio-molecules or a large number of molecules the semi-empirical methods are often used, like CNDO, MNDO, MINDO, AM1 or PM approximations. The idea of this approximation is to replace the complicated electronic potential with empirical parameters. The second quantum chemical method is density functional approximation, which follows different philosophy. Here, the electrons are represented with a cloud of electron density, which is calculated directly using the Kohn–Sham equation [12]. The mostly used results of quantum chemical calculations are orbital energies particularly the highest occupied molecular orbital (HOMO) and the lowest unoccupied molecular orbital (LUMO). Energies (E) of HOMO and LUMO are over Koopmas' theorem related to ionization potential and electron affinity, respectively.

$$IP = -E_{HOMO} \text{ and } EA = -E_{LUMO}$$

They are basis for calculation of two further descriptors: Mullikan electronegativity (ME) and the electronic hardness (N) defined as:

$$ME = (IP + EA)/2 \text{ and } N = (IP - EA)/2.$$

Additionally, the lower and higher molecular orbital energies can be taken as descriptors [13–16]. The quantum chemistry can be used to calculate the charge on particular atoms and the charge distribution in a molecule expressed with the multipole moments (dipole moment, quadrupole moment, etc.). These are often used as descriptors. Often is taken the maximal or minimal charge on particular atom, which is regarded as important in mechanism of activity. Alternatively, the numbers appearing during the quantum chemical calculation, like the electron–electron repulsion energy, or two-electron integrals, can be taken as descriptors [15]. At the end, it is to emphasize that the quantum chemical results depend on the method of approximation and thus it is recommended that all molecules used in a QSAR model are treated within the same quantum chemical approximation.

2.5   Three-Dimensional (3D) Descriptors—Geometrical Descriptors

They are deduced from 3D structure, which is given with positions of all constitutive atoms of a molecule. In the crystalline form of material, the 3D structure is fixed and can be measured with X-ray diffraction measurements. On the other hand, when the molecules are in gaseous phase, in solution, or in the environment of proteins the 3D structures are (usually) flexible. In QSAR studies the 3D structures are mostly determined theoretically by applying of molecular mechanics or quantum chemical methods to minimize the total energy. To this class of descriptors belong: mass distribution descriptors like moments of inertia and gravitation index, shape indices, surface area indices and van der Waals indices, etc. [17–20].

2.6 Different Representations—Fragments as Descriptors

Structural fragments can be used as descriptors. In this description a structure is encoded as multi-dimensional vector where the binary vector components indicate the presence or absence of particular fragment. In the pioneer work Free and Wilson represent molecular structures as a sum of constitutional fragments and correlate it with the activity [21]. Basic idea is that a property (activity, toxicity, etc.) is due to particular molecular fragments. The examples from drug research are presented in references [22, 23]. In environmental toxicology the structural alerts can be applied for study of toxical properties, as for example mutagenicity and carcinogenicity [24, 25].

In the past few decades many software suits have been developed to calculate descriptors like Dragon, MolconnZ, POLLY, APProbe, etc. [for details see Chap. 1 of this book]. Recently released software alvaDesc tool calculates almost 6000 structural descriptors, which are ordered into 33 different groups [26, 27]. Among them 166 belong to the group of fingerprint descriptors (fragments).

2.7 Biodescriptors

In the first AOP step the interaction between a chemical and biomolecules (proteins, enzymes, receptors, nucleid acids) plays the crucial role. The information can be extracted from proteomic studies, the protein structure study, or obtained with compuational modeling like docking or molecular dynamics. It is known that the specific biological environment essentially impacts the molecule's biological properties and shall be included into description of molecular structure (bio-descriptors).

As mentioned above, the proteomic studies can shed light on the mechanism of biological action. Under the term 'proteomic' one understands the status of cellular proteins which can be altered when the cell is treated with xenobiotics. One techniques to examine proteomics is two-dimensional electrophoresis which results in two-dimensional proteomic maps. There are different research aims which require numerical representation of proteomic maps [28–30]. One is the searching for a protein or group of proteins, which are mostly affected after the disturbance of the cell [31–34]. In such cases the numerical representations can be accounted as bio-descriptors. They can be, in addition to structural descriptors, used in QSAR models. In other words, the bio-descriptors describe the status of cells (organisms) after the treatment with a chemical [35, 36].

2.8   Docking

In the recent years molecular docking has gained attention in computational drug design and computational toxicology. Basically, the molecular docking is a computational simulation of interaction between receptor and a ligand molecule [37]. The idea from originate from German chemist Emil Fischer who proposed a century ago the 'key and lock' principle to simulate the interaction between a chemical and a protein. With development of computer hardware and software this technique has been drastically improved mostly due the new information on geometrical data of receptors and ligands. The final result of docking is the optimal conformation considering the condition of the minimal free energy. All docking software packages hame special modules for evaluation of optimal conformations and calculation of binding energy. One of the basic requierments in docking is that the predispositions of the receptors and the binding sites are known, wheter their structures were found experimentally by X-rax, NMR or cryo-microscopy. Alternatively, when the structure of the protein is not known, it can be determinated by searching of similar receptors or analogs (i.e., similar receptors from different species). In era of super-computers diverse strategies how to approach the receptor-ligand interaction are possible [38–40].

In our hierarchical view on molecules docking results, i.e., binding energies and scoring functions, can be considered as bio-descriptors and implemented into QSAR models for properties, which are more complex than receptor-ligand interation.

3   Selection and Reduction of Descriptors

Since the number of descriptors can exceed thousands it is necessary to reduce their number, or to find the 'proper' ones. The careful selection of a handful of descriptors can shed light on mechanism of biological activity. In QSAR research, the subjective selelction of useful descriptors by experts is no longer imaginable today. Here, we shortly introduce four strategies for selection/reduction of descriptors, which are commonly used.

3.1   Pair-Wise Correlation Rules

The basis for this analysis is the correlation matrix between pairs of descriptors. The implemented rule is: If the correlation coefficient between two descriptors is above the threshold one of the descriptor is removed. This procedure runs over all pairs. To additionally reduce the number of descriptors other filtering rules can be implemented, i.e., the standard deviation of descriptor value below the threshold, etc. [41].

3.2   Compressing of Data with Kohonen Artificial Neural Networks

The Kohonen artificial neural networks (KANN) represent a basic algorithm in machine learning and artificial intelligence. Mathematically, it is a mapping of objects of a multidimensional space onto two-dimensional map in a way that similar objects are located close to each other. This property makes the KANN a suitable tool for clustering and classification. For porposes of descriptor selection the KANN are constructed in such a way that enables the variable clustering. In the following, one selects one or more representative descriptors from each cluster [42, 43].

3.3   Principal Component Analysis (PCA)

Principal Component Analysis (PCA) is a widely employed technique for analysing multidimensional data. The basic idea of PCA is to reduce the number of variables while retaining the essential information within the dataset. Mathematically, PCA involves a linear transformation of the original variables into new, latent variables. The transformation matrix is derived from the eigenvectors of the variance matrix, which are organized based on their information content. This means that objects are effectively represented in a new space with only a few variables.

The outcomes of PCA are typically presented through three key components. The informational content of individual latent variables, as well as the cumulative informational content, is commonly presented in a scree plot. The objects presented in the latent variables are visualized in score plots. Furthermore, the loadings analysis is employed to assess which original variables exert significant influence on a given latent variable [44–46].

3.4   Genetic Algoritm

The genetic algorithm (GA) is an algorithm, which imitates a natural evolution. It is often used for searching of descriptor combinations, which providers 'the best' QSAR models. Basicaly, the GA mimics the selection processes occurring in the evolution, which at the end selects the most capable individuals. This is achieved with crossover, mutation and selection of chromosomes. Before the running of GA the fitness score must be defined, i.e., the criterion which selects the survival chromosomes. Usualy the fitness score is one of parameters for evaluation of QSAR models (cross-validation correlation coefficient, mean square error, etc.) The GA runs over many times repeated steps described below.

(i) The population of chromosomes, i.e., manifolds of combinations of descriptors are generated. For all combinations the models are calculated and evaluated. The best models are selected and reported for further analysis.

(ii) In the next generation the combinations of the descriptors from good models exchange some descriptors to produce new combinations (crossover). In addition, some descriptors are randomly exchanged from the pool of all descriptors (mutations).

(iii) The strps (i) and (ii) are repeated until the selected models fulfill the given criteria, or up to the given number of cycles.

For the modeling the different techniques can be applied, like multiple linear regression [41]. Alternatively, for example, references [47–49] describe the counterpropagation artificial neural networks (CP-ANN) as modeling technique and the leave-one-out validation correlation coefficient as scoring function.

4   Structural Similarity

The structural descriptors can be alternatively used to express the similarity between chemical structures and thus a tool for grouping of chemicals, for building of categories and read across. The simplest way to express the similarity is to group the chemicals on the basis of Euklidian distances or inner product between descriptor vectors. More advanced expression of similarity is described by Floris et al. [50]. The similarity between structures is expressed as a similarity index, which is calculated as a weighted sum of factors including molecular fingerprints, constitutional descriptors, heteroatom descriptors and function group descriptors. The weights were optimized in a way to describe the similarity on two large datasets, namely, the set of chemicals with known bio-concentration for fish and log P. The similarity index is incorporated into CAESAR models in a way that the predictions are accompanied with the six most similar compounds from the training set [51].

5   Conclusions

The question of how to represent or encode a chemical structure in a numerical way is the central one in chemical informatics. A representation should be: unique, uniform, reversible, and invariant on translation and rotation of structure. Unique means that different structures have different representations, uniform means that the representation has the same dimension for all structures; reversibility means that the structure can be reconstructed from representation, and invariance means that the representation should not be influenced if the structure is rotated or translated in space. No representation fulfills all of the four requirements simultaneously. For example, in the basic geometrical representation when a molecule is represented with coordinates of all atoms is unique and reversible, but not uniform and invariant. The representation of a structure with a set of descriptors is unique, which means that the descriptors are different for different structures. (This is true only in limited cases. In a special case some descriptors can have the same values for different structures, e.g., isospectral molecular graphs.). In most cases the representation is uniform, what it means is that the same number of descriptors is calculated for all structures. It is not reversible, i.e. the structure cannot be determined from descriptors. It is invariant, which means the orientation of molecules in space does not influence descriptor values.

At the 2D level we describe the structure with atoms and bonds between them ('structural formulas'). At 3D level we describe the structure with positions of all atoms. The step from 2 to 3D description represents a problem. Rigid molecules are rare, 3D structure can be different for molecules in crystalline form, in solution, in gas phase, or in environment of proteins. Often 3D structures are determined theoretically and they are different when the different theoretical approximations are applied. In the determination of 3D structure it is often assumed that the molecules are isolated (in vacuo). In reality the molecules are incorporated into the environment, which in biological systems often consists of bio-molecules. The optimization of 3D structures in the environment of biomolecules is sometimes referred to as 4-dimensionional representation.

Currently available technologies can compute a huge number of chemodescriptors and biodesciriptors for a single molecule or biomolecule. But the critical problem we face is: How to analyze such massive data to derive useful models whereby a handful of selelcted descriptors (independent variables) of our analyses, carried out by robust statistical and/or machine learniing methods, can predict various properties (dependent variables) of chemicals of practical societal interest and produce new knowledge to be implemented in decision support systems. This remains the paramount challenge for data analytics scientists today.

References

1. Ankley GT, Bennett RS, Erickson RJ, Hoff DJ, Hornung MW, Johnson RD, Mount DR, Nichols JW, Russom CL, Schmieder PK, Serrano JA, Tietge JE, Villeneuve DL (2010) Adverse outcome pathways: a conceptual framework to support ecotoxicology research and risk assessment. Environ Toxicol Chem 29:730–741
2. AOP-wiki. https://aopwiki.org/. Accessed 28 Feb 2023
3. Hansh C, Maloney PP, Fujita T, Muir RM (1962) Correlation of biological activity of phenoxyacetic acids with Hammett substituent constants and partition coefficients. Nature 194:178–180
4. Randič M (1975) On characterization of molecular branching. J Am Chem Soc 97:6609–6615
5. Kier LB, Hall LH (1976) Molecular connectivity. Part 7. Specific treatment of heteroatoms. J Phar Sci 65: 1806–1809

6. Balasubramanian K, Basak SC (1998) Characterisation of isospectral graphs using graph invariants and derived ortogonal parameters. J Chem Inf Comput Sci 38(3):367–373
7. Basak SC (1987) Use of molecular complexity indices in predictive pharmacology and toxicology: a QSAR approach. Med Sci Res 15(11):605–609
8. Natarajan R, Basak SC (2011) Numerical descriptors for the characterization of chiral compounds and their applications in modeling biological and toxicological activities. Curr Topics Med Chem 11(7):771–787
9. Natarajan R, Basak SC, Neumann ST (2007) Novel approach for the numerical characterization of molecular chirality. J Chem Inf Model 47:771–775
10. Balaban AT (2001) A personal view about topological indices for QSAR/QSPR. In: QSAPR/QSAR studied by molecular descriptors. Ed- Diudea, M. V., Nova Science Publisher, Inc., Huntington, New York, pp 1–31
11. Roy K, Gosh G (2005) QSTR with extended topochemical atom indices. Part 5: modeling of the acute toxicity of phenylsulfonyl carboxylates to Vibrio fischeri using genetic function approximation. Bioorg Med Chem 13: 1185–1194
12. Lewars EG (2011) Computational chemistry, introduction to the theory and applications of molecular and quantum mechanics, 2nd edn. Springer, Dordrecht, Heidelberg, London, New York
13. Katritzky AR, Lobanov VS, Karelson M (1996) Quantum chemical descriptors in QSAR/QSPR studies. Chem Rev 96: 1027–1043
14. Vračko M, Szymoszek A, Barbieri P (2004) Structure-mutagenicity study of 12 trimethylimidazopyridine isomers using orbital energies and spectrum-like representation as descriptors. J Chem Inf Comput Sci 44:352–358
15. Girones X, Amat L, Robert D, Carbo-Dorca R (2000) Use of electron-electron repulsion energy as a molecular descriptor in QSAR and QSPR studies. J Comp Aided Mol Des 14:477–485
16. Netzeva TI, Aptula AO, Benfenati E, Cronin MTD, Gini G, Lessigiarska I, Maran U, Vračko M, Schüürmann G (2005) Description of the electronic structure of organic chemicals using semiempirical and ab initio methods for development of toxicological QSARs. J Chem Inf Mod 45:106–114
17. Novič M, Vračko M (2001) Comparison of spectrum-like representation of 3D chemical structure with other representations when used for modelling biological activity. Chemom Intell Lab Syst 59:33–44
18. Schuur JH, Selzer P, Gasteiger J (1996) The coding of three-dimensional structure of molecules by molecular transforms and its application to structure-spectra correlations and studies of biological activity. J Chem Inf Comput Sci 36(2):334–344
19. Vracko M, Mills D, Basak S (2004) Structure-mutagenicity modelling using counter propagation neural network. Environ Toxicol Pharmacol 16:25–36
20. Basak SC, Mills DR, Balaban AT, Gute BD (2001) Prediction of mutagenicity of aromatic and heteroaromatic amines from structure: a hierarchical QSAR approach. J Chem Inf Comput Sci 41:671–678
21. Free SM, Wilson JW (1964) A mathematical contribution to structure-activity study. J Med Chem 7:395–399
22. Ursu O, Oprea TI (2010) Model-free drug-likeness from fragments. J Chem Inf Model 50:1387–1394
23. Catana C (2009) Simple idea to generate fragment and pharmacophore descriptors and their implications in chemical informatics. J Chem Inf Model 49:543–548
24. Benigni R, Bossa C (2008) Structure alerts for carcinogenicity, and the Salmonella assay system: a novel insight through the chemical related databases technology. Mut Res 659:248–261

25. Plošnik A, Vračko M, Sollner Dolenc M (2016) Mutagenic and carcinogenic structural alerts and their mechanisms of action. Arh Hig Rada Toksikol 67: 169–182
26. Mauri A (2020) alvaDESC: a tool to calculate and analyze molecular descriptors and finger-prints, in Ecotoxicological QSARs. In: Roy K (ed) New York, NY, Springer US, pp 801–820
27. Mauri A, Bertola M (2022) Alvascience: a new software suite for the QSAR workflow applied to the blood–brain barrier permeability. Int J Mol Sci 23:12882
28. Randić M, Witzmann F, Vračko M, Basak SC (2001) On characterization of proteomics maps and chemically induced changes in proteomes using matrix invariants: application to peroxi-some proliferators. Med Chem Res 10:456–479
29. Randić M, Novič M, Vračko M, Plavšić D (2010) Study of proteome maps using partial order-ing. J Theor Biol 266:21–28
30. Vračko M, Basak SC (2004) Similarity study of proteomic maps. Intell Lab Syst 70: 33–38
31. Vracko M, Basak SC, Geiss K, Witzmann F (2006) Proteomic maps-toxicity relationship of halocarbons studied with similarity index and genetic algorithm. J Chem Inf Model 46:130–136
32. Basak SC, Vracko M, Witzmann FA (2016) Mathematical nanoproteomics: quantitative char-acterisation of effects of multi-walled carbon nanotubes (MWCNT) and TiO_2 nanobelts (TiO_2-NB) on protein expression patterns in human intestinal cells. Curr Comput Aided Drug Des 12:259–264
33. Drgan V, Panek J, Vračko M, Novič M (2014) Encoding and clustering of proteins in mycobac-terium tuberculosis proteome. Int J Chem Model 6: 377–389
34. Vračko M, Basak SC, Witzmann F (2018) Chemometrical analysis of proteomics data obtained from three cell types treated with multi-walled carbon nanotubes and TiO_2 nanobelts. SAR QSAR Environ Res 29:567–577
35. Gute BD, Balasubramanian K, Geiss KT, Hawkins DM (2004) Chemodescriptors versus biode-scriptors for toxicity predictions on halocarbons. Environ Toxicol Pharmacol 16:121–129
36. Basak SC (2011) Role of mathematical chemodescriptors and proteomics-based biodescriptors in drug discovery. Drug Develop Res 72:225–233
37. Fan J, Fu A, Zhang L (2019) Progress in molecular docking. Quantit Biol 7:83–89
38. Lagares LM, Minovski N, Caballero AY, Benfenati E, Wellens S, Culot M, Gosselet F, Novic M (2020) Homology modeling of the human p-glycoprotein (abcb1) and insights into ligand binding through molecular docking studies. Int J Mol Sci 21:4058
39. Bissantz C, Folkers G, Rognan D (2000) Protein-based virtual screening of chemical databases. 1. Evaluation of different docking/scoring combinations. J Med Chem 43: 4759–4767
40. Kokot M, Weiss M, Zdovc I, Anderluh M, Hrast M, Nikolovski N (2022) Diminishing hERG inhibitory activity of aminopiperidine-naphthyridine linked NBTI antibacterials by structural and physicochemical optimizations. Bioorg Chem 128:106087
41. Gramatica P, Chirico N, Papa E, Cassani S, Kovarich S (2013) QSARINS: a new software for the development, analysis, and validation of QSAR MLR Models. J Comput Chem 34:2121–2132
42. Tibaut T, Drgan V, Novič M (2018) Application of SAR methods toward inhibition of bacterial peptidoglycan metabolizing enzymes. J Chemometrics 32:e3007
43. Jezierska A, Vracko M, Basak SC (2004) Counter-propagation artificial neural network as a tool for the independent variable selection: structure-mutagenicity study on aromatic amines. Mol Divers 8:371–377
44. Basak SC, Natarajan R, Mills D, Hawkins DM, Kraker JJ (2005) Quantitative structure-activity relationship modeling of insect juvenile hormone activity of 2,4-dienoates using computed molecular descriptors. SAR & QSAR Environ Res 16:581–606
45. Stanojević M, Vračko M, Sollner Dolenc M (2023) Development of in silico classification models for binding affinity to the glucocorticoid receptor. Chemosphere 336: 139147

46. Stanojević M, Sollner Dolenc M, Vračko M (2023) Predictive models for compound binding to androgen and estrogen receptors based on counter propagation artificial neural networks. Toxics 11: 486
47. Roncaglioni A, Novič M, Vračko M, Benfenati E (2004) Classification of potential endocrine disrupters on the basis of molecular structure using a nonlinear modeling method. J Chem Inf Comput Sci 44:300–309
48. Boriani E, Spreafico M, Benfenati E, Novič M (2007) Structural features of diverse ligands influencing binding affinities to Estrogen α and Estrogen β receptors. Part I: molecular descriptors calculated from minimal energy conformation of isolated ligands. Mol Divers 11: 153–169
49. Bajželj B, Drgan V (2020) Hepatotoxicity modeling using counter-propagation artificial neural networks: handling an imbalanced classification problem. Molecules 25:481
50. Floris M, Manganaro A, Nicolotti O, Medda R, Mangiatordi GF, Benfenati E (2014) A generalizable definition of chemical similarity for read-across. J Cheminform 6:39
51. Vračko M, Drgan V (2017) Grouping of CoMPARA data with respect to compounds from the carcinogenic potency database. SAR & QSAR Environ Res 28:801–813

Chirality Descriptors for Numerical Characterization of Enantiomers and Diastereomers

Ramanathan Natarajan, Subhash C. Basak, and Claudiu N. Lungu

Abstract

There is an increase in the number of chiral molecules coming into pharmaceutical, agrochemical and fragrance industry and this is due to the advancement in chiral synthetic methods using biocatalysts, and separation technology. However, the impetus in the synthesis of enantiopure drugs may partly be attributed to regulatory policies for chiral pharmaceuticals. Enantiomers and diastereomers collectively the stereoisomers have difference in bio-efficacy, adsorption, metabolism, degradation, and toxicity. In the case of achiral molecules, most of these properties have been modelled by quantitative structure–activity relationship QSAR approach using molecular descriptors that encode various features of the molecules. However, extension of this approach to stereoisomers is not possible with the conventional molecular descriptors as these descriptors have the same numerical values for stereoisomers, that is the conventional molecular descriptors are incapable of differentiating stereoisomers. In order to overcome this limitation several attempts have been taken in developing descriptors that differentiate stereoisomers namely enantiomers and diastereomers. Chirality

R. Natarajan
Saranathan College of Engineering, Panjappur, Tiruchirappalli, Tamil Nadu 620012, India
e-mail: natarajan-rd@saranathan.ac.in

S. C. Basak (✉)
1802 Stanford Avenue, Duluth, MN 55811, USA
e-mail: sbasak@d.umn.edu

C. N. Lungu
Department of Morphological and Functional Science, University of Medicine and Pharmacy, 800010 Galati, Romania
e-mail: nicolae.lungu@ugal.ro

about carbon atom(s) being the cause for the stereoisomerism these descriptors are mostly called chirality descriptors and the approach is called as numerical characterization of molecular chirality. These descriptors are expected to play a role not only in developing predictive models for the biological and toxicological properties of the chiral molecules but also assist in identifying new molecules for chiral synthesis. This chapter reviews the various chirality descriptors developed to model the bioactivities of chiral molecules.

Keywords

Chirality • Chirality measures • Chirality index • Enantiomers • Diastereomers • Chiral synthesis • Quantitative structure–activity relationship (QSAR) • Quantitative stereochemical structure–activity relationship (QSSAR) • Chiral pharmaceuticals • Chiral agrochemicals

1   Introduction

Isomerism is the phenomenon by which more than one compound have the same chemical formula but different structural formulae. The word "isomer" is derived from the Greek words "*isos*" and "*mesos*" which mean "equal parts". The term isomerism was coined by the Swedish chemist Berzelius in the year 1830. Hence, isomers are chemical compounds that have identical chemical formulae but differ in the arrangement of atoms in the molecule.

The isomeric forms in which organic molecules exist may be classified into two major types, namely, the structural isomers and the stereoisomers. However, this does not include conformers that are the innumerable positions possible due to rotation about carbon–carbon single bonds (conformers differ only in the torsional angles). Further subclasses of these two major types of isomers are shown in Fig. 1. The structural isomers differ in the connectivity of atoms and this difference confers on them differences in physicochemical and biological properties. In the structural isomers at least one bond must be broken and reconnected to make the structures identical. Among stereoisomers, the geometrical isomerism is due to restricted rotation in a molecule. Geometric isomers otherwise called *cis*- and *trans*-isomers (*E/Z*) differ in their physicochemical properties. In the case stereoisomers due to the presence of stereocenter(s) in a molecule there is no difference in the way the atoms are connected (no difference in connectivity) but they differ in their three-dimensional arrangement or disposition. Hence, enantiomers and diastereomers do not differ in their physicochemical properties. Enantiomers are molecules related as the object and the non-superimposable mirror image while diastereomers are stereoisomers that differ in the three-dimensional arrangement but do not have the mirror image relationship.

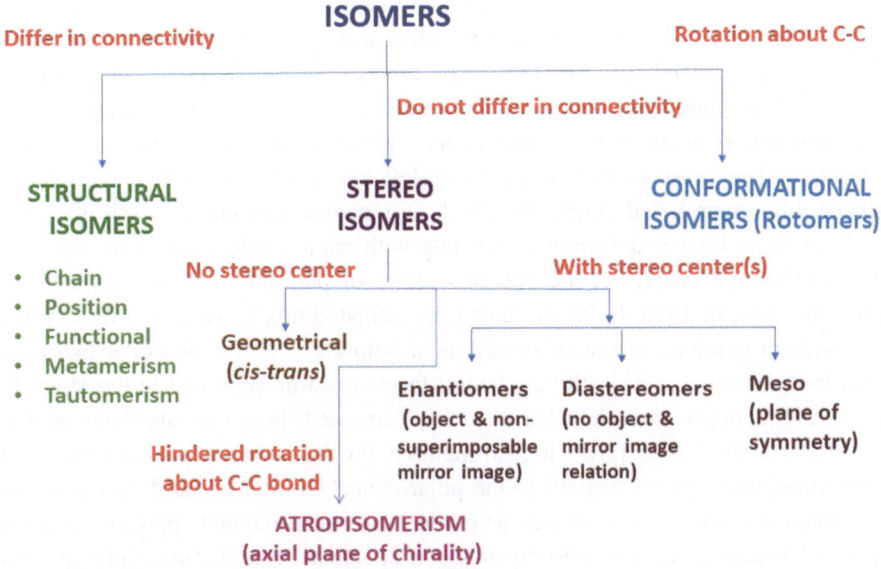

Fig. 1 Classification of isomerism in organic molecules

1.1   Evolution of Chirality

In the early nineteenth century, even though the compounds had a stoichiometric composition of chemical elements, there was only a very little idea of geometric or stereochemical shape of molecules. The notion of shape of chemical compounds began from the theory of Haüy, [1] the French mineralogist and the father of modern crystallography. From his repeated crystal cleavage experiments, he observed the similarity in the external shape of any crystal with the shape of repeating units. However, in 1815, Delafosse and Haüy himself had observed that a few crystals possessing some subsidiary faces were asymmetric which made their morphology and structure to be chiral. In the meantime, Arago [2] discovered the modifications in quartz crystals rotated the plane of polarized light in the opposite direction. These two independent observations by Arago and Haüy were correlated by Sir John Herschel suggesting that the phenomena of opposite rotation are due to hemihedral facets on the crystal. Later, Fresnel explained the optical rotation effect is due to different refractive indices for right- and left-polarized light and postulated that the nature of birefringence or the difference in refractive indices of right- and left-polarized light ($n_L - n_R$) is attributed to their structure. The value of $n_L - n_R$ is positive for dextro-rotatory and negative for laevo-rotatory media [3].

Jean-Baptiste Biot, investigated the phenomenon of optical activity of many natural organic compounds such as sugar, turpentine oil, tartaric acid etc., in their dissolved states at the College de France in 1815. From his observations, he pointed out the difference between optical activity of organic substances and quartz crystal, which loses its crystallinity in dissolved state. In 1832, he analyzed tartaric acid and para-tartaric acid (racemic acid) isolated from Tartar, obtained during wine fermentation and observed that tartaric acid was optically active, whereas the racemic acid later called *rac*-tartaric acid was inactive. The cause of optical inactivity left unexplained by him and Mitscherlich continued this research and concluded that both racemate and enantiomerically pure tartarate have same chemical formulae, crystals with same angles, the same density, and the same refractive index, but the optical activity of para-tartaric acid was different as observed by Biot. In 1848, Louis Pasteur re-investigated this by repeating the crystallization of sodium ammonium salt of tartaric acid below 27 °C. He observed two types of crystals from racemate, one with hemihedral facets towards right and in the other, hemihedral facets laying towards left. He manually separated these crystals (later he devised chemical and biological or physiological methods for the resolution of para-tartaric acid) and examined their optical activity in the polarimeter. Ultimately, he discovered that the optical inactivity of racemic acid was the consequence of equal and opposite deviations of right- and left-handed tartaric acid crystals and concluded that the molecular asymmetry producing non-superimposable mirror images determines the optical activity [4].

In the lectures given by Louis Pasteur before the Council of the Société´ Chimique of Paris in 1860, he described that the organic molecules possess molecular asymmetry on its own even if the physical structure is destroyed, i.e., the crystal is dissolved or fused, the resulted solution or liquid is optically active towards polarized light [5, 6]. Recently [7] had given a comprehensive write up on Pasteur's work on chirality with the list of original references.

This idea of dissymmetry was extended in 1874 by Jacobus Henricus van't Hoff in Holland, and Joseph Achille Le Bel in France. Each of them presented their research papers explaining the atomic arrangements that could produce molecular asymmetry. Earlier, theories of both the scientists were treated as two different approaches of one theory but later analyses clearly showed that they were two different assumptions to connect chirality and optical activity. Based on the idea of August Kekule on the quadrivalency of carbon, van't Hoff introduced the concept of the asymmetric carbon atom as follows: "When the four affinities of the carbon atom are satisfied by four univalent groups differing among themselves, two and not more than two different tetrahedrons are obtained, one of which is the reflected image of the other, they cannot be superposed; that is, we have to deal with two structural formulas isomeric in space that exhibits two distinct non superimposable hemihedral forms which are responsible for optical activity." He also added that all organic compounds in solution that rotate plane of polarization possess an asymmetric carbon. This concept was illustrated with a number of organic compounds such as lactic acid, aspartic acid, asparagine, malic acid, glutaric acid, tartaric acid, sugars and glucosides, camphor, borneol, and camphoric acid. Le Bel's approach was based on neither the tetrahedral model of carbon nor the fixed valences between the atoms. In his theory, he considered the asymmetry of the entire molecule, not of the individual atoms. The biphenyls, allenes and spiranes derivatives are few examples for optically active compounds without asymmetric carbon but with chiral axis confirms Le Bel's assumption on molecular asymmetry [8–12].

Johannes Wislicenus, a German chemist, had long been investigating the isomers of lactic acid (1863–1873). In 1869, he was convinced that the number of genuine lactic acid isomers exceeded that allowed by the existing structural theory given by van't Hoff, and that therefore, this isomerism could only be explained by an extension of structural formulae to show the arrangement of atoms in space. This was later meant by van't Hoff's famous phrase, "I shall devote most of the rest of this paper for discussing the long history of lactic acid, and showing how Wislicenus was eventually forced to the conclusion that the possibilities of the existing theory were exhausted, and that an extension was necessary" [13].

Even though stereoisomerism and optical activity were the subject of discussion in the middle of nineteenth century, none of the scientists who worked in this field used the word 'chirality'. The words 'chiral' and 'chirality' were first used by William Thompson, Lord Kelvin, in an article entitled 'Note on Homochiral and Heterochiral Similarity' to the Royal Society of Edinburgh. He gave the definition for 'chirality' in the footnote of the Second Robert Boyle lecture on the topic "The Molecular tactics of a crystal" to the Oxford university junior scientific club as: "I call any geometrical figure, or group of points, chiral, and say that it has chirality, if its image in a plane mirror, ideally realized, cannot be brought to coincide with itself." This word was derived from the Greek word, "*kheir*," which means "hand", the human hands (or feet) are 'identical opposites,' mirror-images of each other. This is what Pasteur meant by his undefined words 'dissymetrie,' or "dissymetrie moleculaire' [3, 6, 11].

In the twentieth century, the Dutchman Bijvoet determined the actual arrangement in space of the atoms of the sodium rubidium salt of (*b*)- tartaric acid using X-ray diffraction technique (exactly 100 years before, the first resolution of *d, l*-tartaric acid was done by Pasteur) and thus made the first determination of the absolute configuration about an asymmetric carbon [6, 14].

1.2   Need for New Chiral Descriptors

Enantiomers are differentiated only in a chiral environment. Biological systems always have chiral environment because their building blocks are chiral molecules such as sugars, α-amino acids and nucleic acids. Hence, stereoisomers may differ in their pharmacokinetic, pharmacodynamic and toxicological properties because there would be selective absorption, protein binding (ligand-target interaction), transport, metabolism etc. Biological receptors, enzymes, and ion channels are chiral in nature and their stereospecificity for chiral ligands is quite striking. The difference in the bioactivity of enantiomers of

Table 1 Importance of chirality in various therapeutic agents

No	Therapeutic purpose	References
1	Ocular agents	Leonov and Bielory [17]
2	Antiallergic and immunologic drugs	Bielory and Leonov [18]
3	Antidepressants and psychiatry drugs	Baker and Prior [19], Lane and Bake [20], Budău et al. [21], Howland [22], Nageswara Rao and Guru Prasad [23]
4	Antimicrobial	Hutt and O'Grady [24]
5	Antirheumatic drugs	Kean et al. [25]
6	Antiarrhythmic drugs	Mehvar et al. [26]
7	Cardiovascular drugs	Ranade and Somberg [27]
8	Antiasthma agents	Vakily et al. [28]
9	Anticancer drugs	Wainer and Granvil [29]
10	Drugs for arthritic disorder	Williams [30]
11	Anesthesia	Nau and Strichartz [31], Čižmáriková et al. [32]

various therapeutic agents have been studied and reported (Table 1). The importance of chirality in agrochemicals is documented in the monograph by Kurihara and Miyamoto [15]. Differential biological recognition of stereoisomers is very striking among insects and especially their olfactory receptors are highly specific in differentiating pheromones, kairomones, and chemical cues from hosts [16].

The phenomenon of chirality will continue to dominate the pharmaceutical and agrochemical industry and thus offer new challenges. Hence, it is necessary to come up with toxicological prediction for not only the chiral isomers but also for the new chiral molecules so that they are "benign by design". Prediction of pharmacological and toxicological properties of new molecules even before they are synthesized is possible from the predictive models built using Quantitative structure–activity relationship (QSAR) approaches wherein calculated molecular descriptors, and quantum chemical parameters calculated in vacuum from molecular structure alone and without the use of any other experimental data are applied. All these computed descriptors fail to distinguish enantiomers and diastereomers. Structure–activity relationship models of biological and toxicological properties of stereoisomers need special type of molecular descriptors that include the information about the chiral center(s). There are several attempts in developing descriptors that encode information about chirality in the molecule. One of the earliest reviews of chirality measures was by Buda et al. [33] while the review by Weinberg and Mislow [34] discussed chirality measures reported until 1995. In these reviews, chirality is treated more as a domain of mathematics and chirality measures of shapes or objects dominated rather than chiral molecules and stereoisomerism. The review by [35] covered both chirality and symmetry measures. Crippen [36] compiled the various methods of converting conventional molecular descriptors or invariants of molecular graphs into chirality descriptors by including a chiral correction. He treated the chirality descriptors both as qualitative and quantitative measures. The extended molecular descriptors are qualitative descriptors while he classified the computation of quantitative chirality descriptors into three approaches. The first approach is to use van der Waals volume overlap of a molecule and its mirror image and the second one is to measure the degree of distortion needed to convert the molecule or its subset of atoms into a structure of desired symmetry. The third approach involves translation and rotation about a symmetry axis to match the various atomic properties. Natarajan and Basak [37, 38] not only reviewed the various chirality descriptors but also their application in modelling biological activities of enantiomers and stereoisomers. This article discusses the early attempts briefly while the molecular chirality descriptors, their calculation and applications are dealt with in detail.

2 Chirality Descriptors

2.1 Asymmetry Product

Asymmetry product is one of the oldest chirality measures introduced by [39]. It is a continuous function of a set of chiral molecules but vanishes at achirality. The set is assumed to be a regular tetrahedron with the four ligands representing the four vertices. The asymmetry product P_A is then,

$$P_A = \prod_{i=1}^{6} d_i \tag{1}$$

where d_i are the distances from the mass (molecular) center to the six planes of symmetry of the regular tetrahedron (no change in the bond angle from 109° 28′ is allowed). Asymmetry products were correlated with optical rotatory power of optically active compounds. If the masses of the four ligands (groups) attached to the asymmetric carbon are m_1, m_2, m_3 and m_4 then asymmetry product based on mass differences P_m will be proportional to the asymmetric product P_A based on distances.

$$P_m = \prod_{i>j}^{1-4} (m_i - m_j) \tag{2}$$

Brown [40] generalized the asymmetric product where the mass was replaced with any appropriate property of the four ligands. In 1934 Boys [41, 42] proposed a similar equation to calculate the optical rotatory power of chiral molecules. Nearly after three decades Ruch [43–45] generalized asymmetry product as "chirality function." Generalized chirality functions and their relations with group theory were discussed by several authors and there were controversies about ligand properties and the limitations in using only in a predefined class such as pyramidal, bipyramidal etc. [35].

2.2   Chirality Measures Based on Hasudorff Distances

An alternative method to compare shapes or molecules by one-to-one correspondence or mapping is the quantification of similarity using Hasudorff distance between two arbitrary point sets. Rassat [46] suggested the Hasudorff distance as a similarity measure to define the left–right classification. Hasudorff distances between enantiomers (mirror image objects) was used by [47] as a similarity measure. They used the minimized distances and they were normalized using the highest Hasudorff distance between two points of the set. Mezey [48] extended the idea of Rassat by using molecular electron density clouds that is, he viewed the ligands as electron clouds of atoms in combination rather than mere atoms. The Hasudorff distances calculated using fuzzy theory, fuzzy Hasudorff-type distances between iso-density electron contours were used to compute the chirality measures. A series of chirality measures were developed by constructing the topological symmetry trees [49, 50]. Following this, Mezey published a series of articles on quantum similarity [51], fuzzy chirality measure and generalized chirality and symmetry measures [48]. Following this [52] came up with a new quantum chemical-based chirality measure that uses electronic wave function and its inverted image. Luzanov et al. [53, 54] proposed a κ-index from the spatial curves of electron paths in molecules and in 2007 [55] a one-electron invariant is proposed as a chirality measure.

2.3   The *Aufbau* Approach

Aufbau is a German word means building up and the word is not new to chemists because they use *aufbau princ*iple in filling up atomic orbitals with electrons. The approach that uses adding up features of chirality in calculating chirality measures is thus called the *aufbau* approach. The 3-D chemical structure of a chiral molecule can be represented in 2-D by Fischer projections. For each enantiomer there are 12 possible Fischer projections (Fig. 2) and thus 24 of them are possible. Each of the Fischer projection might be represented by a matrix. Capozziello and Lattanzi [56] applied an algebraic approach wherein they used matrix operators based on $O(4)$ orthogonal group for each of the 24 matrices to obtain a three-way classification of achiral and chiral. They proposed a chirality index χ [57] by projecting the four ligands around a chiral atom on a $\{x; y\}$-plane. The chirality index χ was obtained using complex numbers to define the bond length and bond angles (angular positions of the ligands). The approach was extended as a building up (*aufbau*) procedure to predict the chiral features of a new molecule by "adding up" [58]. The chirality index $\chi \equiv \{n,p\}$; where n is the number of stereo-genic centers and p the number of inversions (at most one for each center). In this approach, adding up a chiral center to the structure gives rise to a new molecule, where $\chi \equiv \{n + 1, p + \Delta p\}$, the variation of p, Δp ($\Delta p = 0,1$) gives the chiral feature of the new molecule. In the case of a *meso* compound the added chiral center is not different from the first chiral center and hence, the *aufbau* approach failed in such cases.

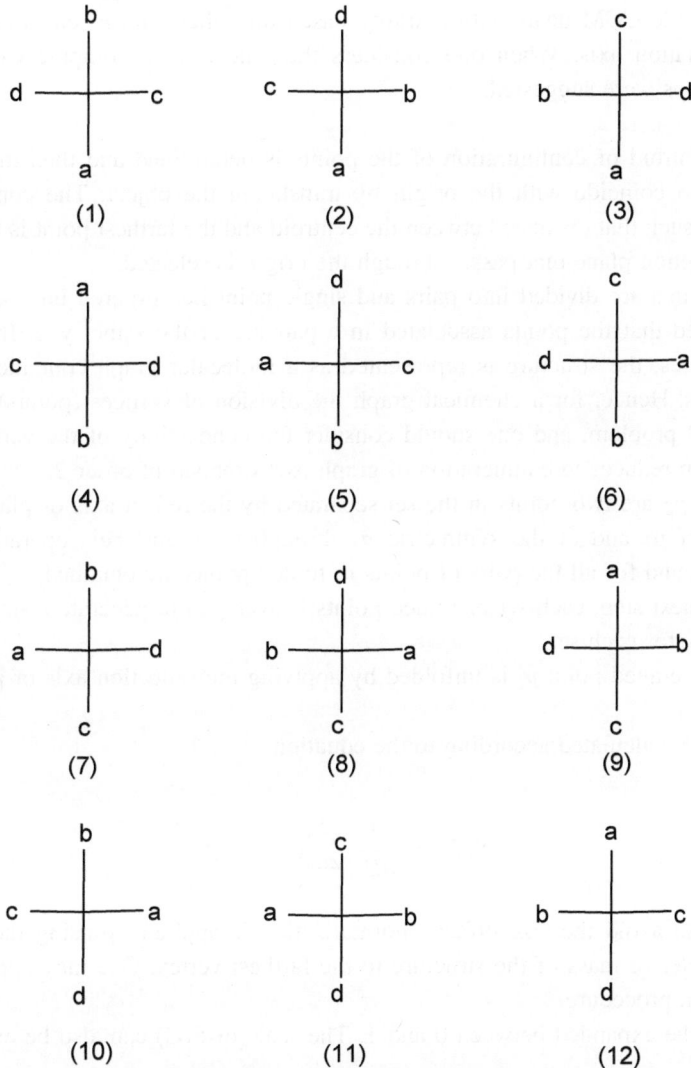

Fig. 2 The twelve possible Fischer projections for an enantiomer

2.4 Continuous Chirality Measure

In 1992 Zabrodsky et al. [59] introduced Continuous Symmetry Measures (CSM) and Continuous Chirality Measures (CCM). As the very names suggests CSM deals with symmetry while CCM deals with chirality based on either with a reflection or with an improper rotation axis. When one considers the reflection to compute CCM then the following steps were suggested:

(i) The centroid of configuration of the points is determined and then the centroid is made to coincide with the origin by translating the object. The configuration is scaled such that distance between the centroid and the farthest point is taken as one.

(ii) A reflection plane that passes through the origin is selected.

(iii) The points are divided into pairs and single point i.e., ordered into sets. It should be noted that the points associated in a pair are of the same type. In the case of molecules, the structure is represented as a molecular graph contained edges and vertices. Hence, for a chemical graph the division of vertices (points) is a combinatorial problem, and one should consider the connectivity of the vertices and the problem reduces to enumeration of graph isomorphism of order 2.

(iv) p_1 and p_2 are two points in the set separated by the reflect axis or plan σ the midpoint of p_1 and of the symmetric p_2 through the plane This operation is called folding and for all the pairs of points p_i folded points are obtained.

(v) In the next step, each set of folded points is averaged to generate a single averaged point $\widehat{p}_i$ for each set.

(vi) Each averaged point $\widehat{p}_i$ is unfolded by applying the reflection axis or plane σ as in step (iv).

(vii) $S(S_{2n})$ is calculated according to the equation:

$$S(G) = \frac{1}{nD^2} \sum_{i=1}^{n} (p_i - \widehat{p}_i)^2 \tag{3}$$

In order to avoid the size effects, normalization is applied by using the distance D from the center of mass of the structure to the farthest vertex. One may apply any other normalization procedure.

S(G) can be expanded between 0 and 1. The scale of S(G) can also be expanded to 0 to 100 as

$$S^1 = S(G) \times 100$$

Calculation of CCM associated with improper axis of rotation of even order, S_{2n} involves the following steps:

(i) The set points i.e., the configuration is normalized as in step 1 and 2 for CCM reflection plan.

(ii) In this CCM measure an improper axis passing through the origin is selected.

(iii) The points are divided into ordered pair of points and single points. In the case of a set containing single points that point is multiplied 2n times while the set two points each of the two points is multiplied n times.

(iv) Folding of each set of points is carried out using rotations about the improper axis is followed by inversion. Thus, the folded points $\hat{p}_i$ are obtained by applying rotation—inversion.

(v) A single averaged point $\hat{p}_i$ is obtained for each set of folded points.

(vi) Each of the average point $\hat{p}_i$ is unfolded by applying reverse rotations.

(vii) The sum S of the n squared distances is calculated as mentioned in the calculation of CCM for reflection.

The above steps (ii)–(vi) are repeated with all possible division of points into sets for all improper rotation axis to minimize S. (In the case of molecules, especially organic molecular chirality is associated with an asymmetric or chiral carbon that is linked to four different atoms as CWXYZ). While most of the approaches in numerical characterization of chirality treat chirality as a binary (+1 or −1) CCM considers chirality as a continuous structural property. That is chirality is treated on a grey scale instead of a black-or-white scale.

Initially [60] CCM and CSM were applied to several shapes such as triangles, tetra-hedrons etc., to identify the most chiral objects. Then they applied to predict chirality change during conversion of one enantiomer to the other (enantiomerization pathway), diastereomerism and prochirality, chiral C_{28} fullerene, helicenes, chiral diffusion-limited aggregates (DLA). Keinan and Avnir [61, 62] applied CCM to study recognition of substrate by receptors for (1) trypsin/arylammonium inhibitors, (2) the D_2-dopamine receptor/dopamine derivative agonists, (3) trypsin/organophosphate inhibitors, (4) acetyl-cholinesterase/ organophosphates and (5) butyrylcholinesterase/ organophosphates. The structures of inhibitors are given in Fig. 3. The research group developed an online tool for computing CCM and CSM [63, 64]

2.5   Molecular Similarity Approach

One of the simplest methods to measure molecular similarity is the one-to-one mapping of the structure [65, 66]. Seri-Levy et al. generated similarity indices by superimposing closely related structures. Different physiochemical properties such as electron density, electric field, elements of symmetry, electrostatic potential (ESP), and shape could be used as the basis for computing similarity indices. In the case of ESP, the similarity index R_{AB} for the superimposed molecules A and B can be obtained from Eq. 4.

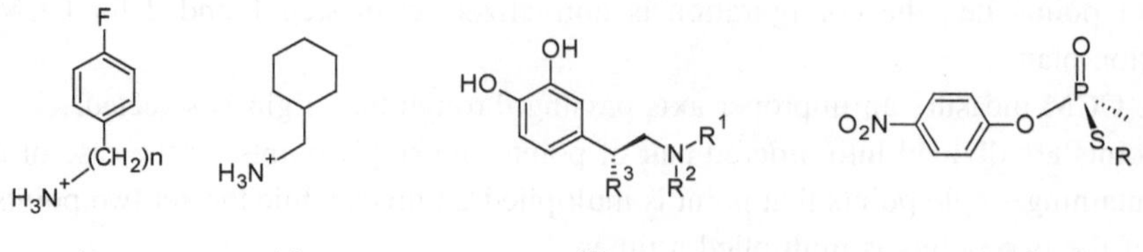

(1)   (2)   (3)   (4)

Dopamine derivatives   Organophosphates

($R^1 = R^2 = R^3$ = H: Dopamine)   R = CH_3, C_2H_5 n-C_3H_7,n-C_4H_9, n-C_5H_{11}

Fig. 3 The four classes of substrates used to model the substrate-receptor recognition

$$R_{AB} = \left(1 + \frac{\int P_A P_B d\gamma}{\left(\int P_A^2 d\gamma\right)^{0.5}\left(\int P_B^2 d\gamma\right)^{0.5}}\right)/2 \tag{4}$$

where P_A and P_B are the ESPs at a point in space.

The similarity index defined in Eq. 4 is normalized and has values between 0 and 1. Similarity index based on shape that is, the shape similarity index S_{AB} is defined by Eq. 5.

$$S_{AB} = \frac{C}{(T_A T_B)^{0.5}} \tag{5}$$

For computing shape similarity index S_{AB}, van der Waals volumes were used, and the molecules were enclosed in a three-dimensional grid box. T_A and T_B are the number of grid points and C is the number of grid points falling inside both the enantiomers. Similarity indices thus generated were correlated with eudismic ratio of 3-hydroxyphenylpieridines (**5**). In a similar approach [67] used the principal components of the complete $N \times N$ pairwise similarity matrices. Calcium channel antagonistic property of dihydropyridines (**6**) was studied to order the eudismic activity of the enantiomers (Fig. 4).

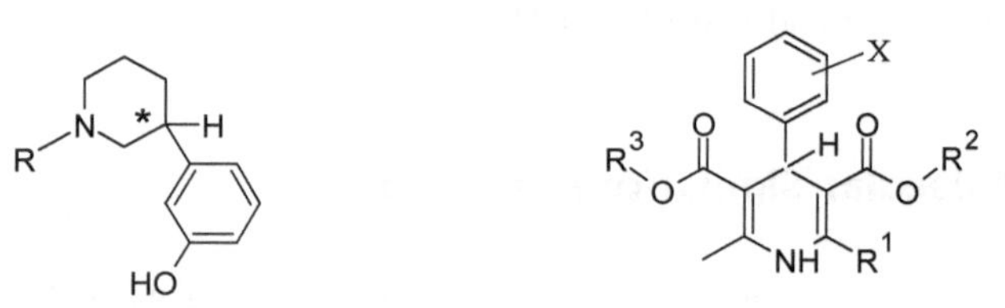

(5)   (6)

3-hydroxyphenylpiperidines (3HPP)   Dihydropyridines

Fig. 4 Structure of 3HPP and dihydropyridines

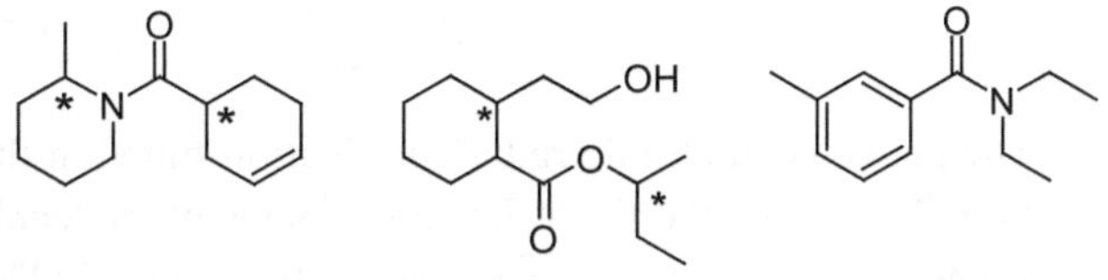

Fig. 5 Structures of the insect repellents DEET, SS220 and picaridin

(7)   (8)   (9)

AI3-37220   Picaridin   DEET

The mapping may be based on the atom type [68, 69]. Basak et al. [68] used molecular overlay wherein optimized geometries were overlaid and the similarity index was measured as the root mean square distance (RMSD) between the corresponding atoms. Molecular overlay approach was used by them to rank the repellency of diastereomers of the insect repellents AI3-37220 (**7**) and picaridin (8). In their study they used DEET (**9**) as the reference (gold standard) repellent and all the other molecules were overlaid on the optimized geometry of DEET molecule (Fig. 5).

2.6   Chirality Codes

Several numerical descriptors described in the previous sections are based on similarity measures. One of the early numerical characterizations that found application in modeling properties of chiral molecules is 'chirality code' developed by Aires-de-Sousa et al. [70, 71], Zhang et al. [72]. For the sake of creating chirality code 3D structure of a molecule is converted into a radial distribution function. The atomic property a may be atomic number, partial atomic charge, atomic polarizability etc. The new parameters E_{ijkl} calculated as per Eq. 6 does not contain any information about the stereochemistry.

$$E_{ijkl} = \frac{a_i a_j}{r_{ij}} + \frac{a_i a_k}{r_{iik}} + \frac{a_i a_l}{r_{il}} + \frac{a_j a_k}{r_{jk}} + \frac{a_j a_l}{r_{jl}} + \frac{a_k a_l}{r_{kl}} \tag{6}$$

where, i, j, k, l are four atoms and a is the atomic property.

r_{ij} is the shortest through-bond distance and the atomic property a may be atomic number, partial atomic charge, atomic polarizability etc.

E_{ijkl} thus calculated does not encode any information about molecular stereochemistry. The information regarding stereochemistry is introduced using another quantity called chirality signal C_{ijkl}. Chirality signal is calculated by ranking the four atoms based on the atomic property i and the neighbors are also considered for ranking. The ranking is similar to the assignment of R and S configurations following the Cahn, Ingold and Prelog (CIP) system. If the ranking is in clockwise direction, of course with the fourth atom (lowest ranked) behind the plane, the chirality signal gets a value of $+1$ and for ranking in counterclockwise the value is -1. The two values E and C are calculated for all the combinations of the four atoms. Then the chirality code is:

$$f_{cc}(x) = \sum\nolimits_{i,j,k,l} C_{ijkl} e^{-B\{x-E_{ijkl}\}} \tag{7}$$

The above is an atomic radial distribution function that contains stereochemical information. To obtains the Cartesian coordinates of molecules the software CORNIA™ was used and the 3-D structures were generated using PETRA (Parameter Estimation for the Treatment of Reactivity Applications) (molecular-networks.com, Software/PETRA/). Chirality codes were used in prediction of enantioselectivity in chiral chromatography [73], enantioselectivity of reactions [71, 74, 75] and prediction of NMR chemical shifts and absolute configuration from NMR [72].

2.7   Physicochemical Atomic Stereo-Descriptors (PAS)

In continuing their work on chiral descriptors [76] introduced the physicochemical atomic stereo-descriptor (PAS). The enantiomers and diastereomers were classified as R and S based on the priorities assigned to the four ligands attached to a chiral center. However, the priority was not based on the Cahn, Ingold and Prelog (CIP) rules but based on the physicochemical properties of the four ligands. Twenty-one physicochemical properties of the ligands attached to a chiral center were considered. Hence, one has two types of classification namely, (1) based on the CIP nomenclature and (2) CIP-like method used by Aires-de-Sousa et.al. If a chiral descriptor is computed after ranking the four ligands according to the CIP rules, then the descriptor is called CIP or R/S descriptor. On the other hand, CIP-like or R/S-like chirality descriptors are obtained when physicochemical properties of the four ligands are used for ranking them to assign R/S configuration. This is illustrated for bromo(chloro)fluorobutane (Fig. 6a) and 2-aminobutanoic acid (Fig. 6b). In the case of S- bromo(chloro)fluorobutane, the S-configuration is assigned based on the CIP rules for the same stereoisomer when atomic masses of the ligands are considered the configuration remain as S but it is denoted as S-like on the other hand when electronegativities of the ligands are considered the priority changes F (4.0) > Cl (3.0) > Br (2.8) and therefore it becomes R-like configuration. 2(R)-2-aminobutanoic acid has S-like configuration when the total atomic masses of the ligands (group molecular masses) are considered.

Some of the properties considered to assign R/S-like configuration are: number of atoms in the immediate sphere from the chiral center, number of atoms within the third sphere of bonds from the chiral center, topological distance from the chiral center to the farthest atom in the ligand or the substituent, maximum topological distances between two vertices, partial atomic charges, bond polarizabilities, and resonance stabilization of the charge created due to bod breakage.

Aires-de-Sousa et al. [70] used the same dataset as in their previous studies on chirality codes. The two asymmetric reactions considered for the application of PAS descriptors are:

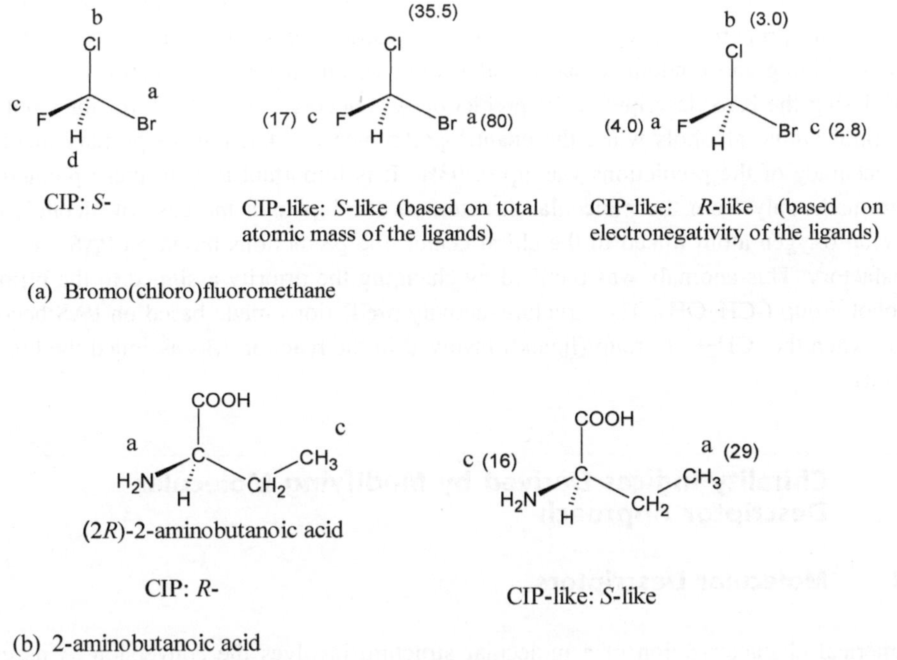

(a) Bromo(chloro)fluoromethane

(b) 2-aminobutanoic acid

Fig. 6 Assignment R/S (CIP) and CIP-like configuration. (In example **a**, for bormo(cholro)fluoromethane *CIP* S-configurations remains CIP-like S when atomic masses are considered for ranking while the same become CIP-like *R* when electronegativities are considered for ranking. In example **b** 2-aminobutanoic acid CIP *R* isomer becomes CIP-like *S* isomer when sum of atomic masses of each of the four ligands are used for ranking)

(1) prediction of major enantiomer in the catalytic addition of diethyl zinc to benzaldehyde using chiral amino alcohols as catalysts (the data set had forty-eight enantiomeric pairs of chiral amino alcohols) (Fig. 7).

(2) prediction of enantioselectivity in the bio-catalysis of eighty-six enantiomeric pairs of primary alcohols by *Pseudomonas cepacia* lipase (PCL).

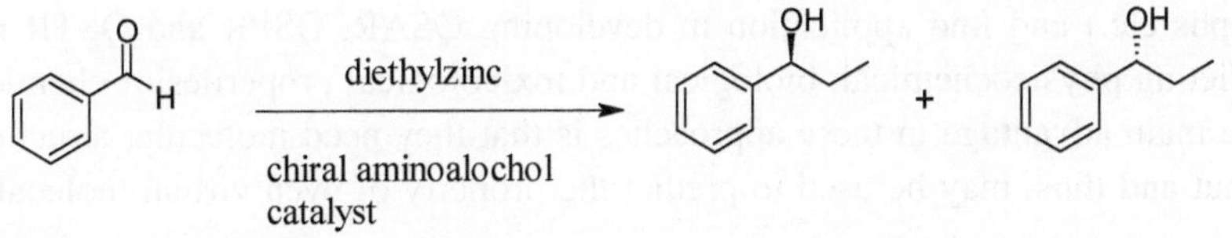

Fig. 7 Reaction between diethyl zinc and benzaldehyde catalyzed by amino alcohol

PETRA (Molecular Networks Gmbh, Erlangen, Germany) was used to compute the various atomic properties. They used counter-propagation neural networks (CPG NN) for model building and random forest tree algorithm was used for classification and prediction. Using the PAS descriptors, the prediction was accurate up to 90% for the catalysis by chiral amino alcohols while the enantio-preference of PCL toward primary alcohols the accuracy of the predictions was up to 93%. It is important to note these predictions were made only from the molecular structure as the input. In the case of alcohols that have an oxygen atom linked to the chiral center, the predictions based on PAS were not satisfactory. This anomaly was rectified by changing the priority assigned to the primary alcohol group (-CH$_2$OH). The structure–activity predictions made based on PAS become good when the -CH$_2$OH group (ligand) involved in the reaction was assigned the highest priority.

3   Chirality Indices Derived by Modifying Molecular Descriptor Approach

3.1   Molecular Descriptors

Numerical characterization of a molecular structure involves the conversion of its constitution into a number or a set of numbers using well-defined algorithms. Thus, the mathematical representation and characterization of a molecule might involve mathematical methods such as graph theory, or matrix and determinant operations. Such operations encode some selected structural features of the molecule and the molecular descriptor generated find applications in quantitative structure–property/ activity relationships (QSPR/QSAR) studies and quantitative molecular similarity studies [77–82]. Many such descriptors are computed using graph-theoretic principle and are regarded as graph invariants. The molecular structure is transformed into a graph called molecular graph. A molecular graph is usually a hydrogen suppressed where the non-hydrogen atoms are the vertices, and the bonds are the edges. A molecular graph is usually an undirected weighted graph. By application of graph theory to the molecular graphs a large number of graph invariants can be derived. These graph invariants are called molecular descriptors and are popularly known as topological indices. However, they do not encode the three-dimensional topology of the molecules. There are quite a large number of molecular descriptors derived from molecular graphs (simple, multi, pseudo, vertex-weighted or edge-weighted molecular graphs etc.) and find application in developing QSAR, QSPR and QSTR modeling for predicting physicochemical, biological and toxicological properties of chemicals [77–87]. The main advantage in these approaches is that they need molecular structure as the only input and thus, may be used to predict the property of even virtual molecules.

3.2　　Chiral Modifications

Most of these graph theoretical descriptors are incapable of discriminating enantiomers/diastereomers of chiral chemicals. Hence, several modifications in their computation have been suggested and these modifications may be classified into two types. (1) Adding chiral correction into the connectivity matrix and computing new standalone chiral descriptors (2) applying a chiral correction factor such that new *Chiral TI = conventional TI ×* *correction factor*. The second approach modifies the existing topological index and therefore, they must be computed as the first step. The chirality descriptors can distinguish the enantiomers as the first requirement and they find application in modeling biological and other properties of the chiral molecules. For several of the chirality descriptors explained in this section certain benchmark data sets comprising of chiral molecules are used for QSAR modeling. These data sets include seven pairs of 3-hydroxyphenyl piperidines 3HPPs (**5**) [88], 31 Cramer's steroids [89, 90], 66 Histamine H1 receptors [91], 9 HIV-1 protease Inhibitors [92, 93], and 78 ecdysteroids [94].

3.3　　Standalone Chiral Descriptors

The first step in deriving a graph invariant from a molecular graph involves the representation of the graph in the form of a matrix. There are a large number of graph theoretical matrices such as vertex matrices, edge matrices and incidence matrices. Among these the adjacency matrix is the simplest and is written based on atom connectivity. The adjacency matrix of a molecular graph G with n vertices is a $n \times n$ square symmetric matrix $A = [a_{ij}]$ such that

$a_{ij} = 1$ if there is an edge (bond) between the ith and the jth vertices (atoms) and $= 0$ if there are no edge (bond) between them

In other words, all elements of an adjacency matrix are zeros except $a_{ij} = a_{ji} = 1$ when the two atoms i and j are connected by a chemical bond.

In the adjacency matrix a chiral correction can be introduced. For the chiral atom in R-configuration the vertex degree is $(a_i + c)$ and for S-configuration the vertex degree becomes $(a_i - c)$. This is equivalent to substituting the main diagonal elements a_{ii} of the adjacency matrix A with $+ c$ or $-c$ for an asymmetric carbon in R- or S- configuration, respectively. The calculation of Zagreb index from chiral modified adjacency matrix is illustrated below for bormo(chloro)fluoromethane (**10**) (Fig. 8).

$$\begin{array}{cccc} 1 & 2 & 3 & 4 \end{array}$$

The adjacency matrix for graph, $\quad A(G) = \begin{bmatrix} 0 & 1 & 0 & 0 \\ 1 & 0 & 1 & 1 \\ 0 & 1 & 0 & 0 \\ 0 & 1 & 0 & 0 \end{bmatrix} \begin{array}{c} 1 \\ 2 \\ 3 \\ 4 \end{array}$

Row sums for $A(G) = 1, 3, 1, 1$

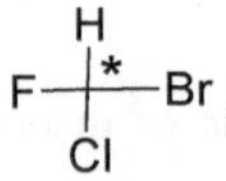

(10)
Bromo(chloro)fluoromethane

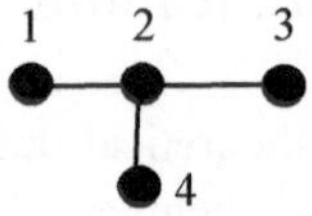

H-supressed moelculr graph (G)

Fig. 8 Molecular structure for bromo(chloro)fluoromethane and its hydrogen suppressed molecular graph

Adjacency matrix with chiral corrections for *R*

$$A(G_R) = \begin{array}{c} \begin{array}{cccc} 1 & 2 & 3 & 4 \end{array} \\ \begin{bmatrix} 0 & 1 & 0 & 0 \\ 1 & \mathbf{+1} & 1 & 1 \\ 0 & 1 & 0 & 0 \\ 0 & 1 & 0 & 0 \end{bmatrix} \end{array} \begin{array}{c} 1 \\ 2 \\ 3 \\ 4 \end{array}$$

Row sums for $A(G_R) = 1, 4, 1, 1$

Adjacency matrix with chiral corrections for $S =$

$$A(G_S) \begin{array}{c} \begin{array}{cccc} 1 & 2 & 3 & 4 \end{array} \\ \begin{bmatrix} 0 & 1 & 0 & 0 \\ 1 & \mathbf{-1} & 1 & 1 \\ 0 & 1 & 0 & 0 \\ 0 & 1 & 0 & 0 \end{bmatrix} \end{array} \begin{array}{c} 1 \\ 2 \\ 3 \\ 4 \end{array}$$

Row sums for $A(G_S) = 1, 2, 1, 1$

$$\text{Zagreb index } M_1 = \sum a_i^2 \tag{8}$$

$$\text{Zagreb index } = 1^2 + 3^2 + 1^2 + 1^2 = 12$$

$$\text{Modified Zagreb index for } R - \text{configuration } = 1^2 + 4^2 + 1^2 + 1^2 = 19$$

$$\text{Modified Zagreb index for } S - \text{configuration } = 1^2 + 2^2 + 1^2 + 1^2 = 7$$

Schultz and co-workers [95] and [96] suggested new chiral descriptors by introducing the chiral correction of $+1$ for a (*R*) carbon or -1 for a (*S*) carbon. Schultz et al. [95] computed the new chiral topological indices via a vertex-weighted distance matrix. While Juilan-Ortiz et al. [96] used the vertex degrees with the chiral correction (+1 or -1) calculated connectivity indices proposed by Hall and Kier [97]. Using the new chirality indices, they developed QSAR model for the seven pairs of 3HPPs for their σ and dopamine D2 receptor activities.

Golbraikh et al. [98] suggested that chirality correction *c* could be a real or imaginary number. The chirality correction was denoted as *ic,* where $i = \sqrt{-1}$, and *c* is always a real number. The vertex degree $a_{ii} = (a_i + ic)$ or $(a_i - ic)$ for an atom in R-configuration or S-configuration, respectively. Hence, two series of chirality descriptors namely, one based on real number chiral correction and the other based on imaginary number chiral correction were introduced. Thus, a series of chirality descriptors were computed. These descriptors were implemented in QSAR modelling and k-nearest neighbor (k-NN) variable selection method was used to select the chiral predictors [99]. The QSAR models were developed [100] for four sets of data comprising of chiral molecules namely, 78 ecdysteroids, 31 Cramer's steroids, 66 histamine H1 receptor ligands and 49 HIV-1 protease inhibitors.

Lukovits and Linert [101] modified the first order connectivity index $^1\chi^v$ using a chiral function that has positive and negative values for D and L enantiomers. That is $F(D) = -F(L)$, where F is the chiral function, D and L denote the two enantiomers with relative configurations D and L, respectively. The new chirality index with the thin-layer chromatographic retention indices of four pairs of hydroxy acids and five pairs of ten α-amino acids [102].

Yang and Zhong [103] came up with chirality factors (CF) that were then used to modify the conventional TI into chiral topological indices using the relation:

$$TI_{CF_n} = TI \times {}^nCF \tag{9}$$

where TI_{CF_n} is a chiral topological index, TI is the conventional topological index and ^nCF is the n-order chirality factor. In order to obtain the chirality factor, the four ligands attached to a chiral center were assigned priority according to the CIP-rule. The lowest priority substituent (D) was placed away from the plane and the other three substituents, which form a plane, were defined as A, B and C clockwise. The chirality factor of nth order was calculated using Eq. (10).

$$^nCF = \frac{N}{M} \sum_{i=1}^{m} \left(\frac{{}^nf_{iA} - {}^nf_{iB}}{{}^nf_{iA}^nf_{iB}} \right) \left(\frac{{}^nf_{iA} - {}^nf_{iC}}{{}^nf_{iA}^nf_{iC}} \right) \left(\frac{{}^nf_{iA} - {}^nf_{iD}}{{}^nf_{iA}^nf_{iD}} \right) \left(\frac{{}^nf_{iB} - {}^nf_{iC}}{{}^nf_{iB}^nf_{iC}} \right) \left(\frac{{}^nf_{iB} - {}^nf_{iD}}{{}^nf_{iB}^nf_{iD}} \right)$$
$$\left(\frac{{}^nf_{iC} - {}^nf_{iD}}{{}^nf_{iC}^nf_{iD}} \right) \tag{10}$$

where N is the number of non-hydrogen atoms, M is the number of chiral centers, $^nf_{iA}$ is the chiral index of substituent A attached to chiral center i. The chiral index ^nf_i is given by Eq. (11).

$$^nf_i = \sum_{i=1}^{m} \frac{\delta_j}{d_{ij}^2} d_{ij} \leq n \tag{11}$$

where m is the number of non-hydrogen atoms in the ligands (A, B, C or D) whose distance to chiral center i is less than or equal to n and the hydrogen atoms directly attached to the chiral center, d_{ij} is the graph distance of vertex j to chiral center i, (the shortest path between the vertex i and j, and δ_j is the degree of vertex j) and this is zero for hydrogen. The chiral correction factor becomes zero for achiral molecules and the chirality topological indices calculated for a pair of enantiomers using this approach has the same numerical values but opposite sign. A closer look at Eq. (10) reveals that it is nothing but a modified form of asymmetry product defined in Eq. (2). Computations of chirality factors considered the importance of distance of an atom from the chiral center and contribution by the lowest priority group. They applied the new chirality indices in the QSAR modeling of 3HPP data set [88], high-pressure thin-layer chromatography (HPTLC) [102], ED_{50} data for 23 ecdysteroids [94], and Cramer's data set [89].

3.4   Stochastic and Non-Stochastic Indices Based on Atom or Bond Adjacency

Diaz et al. [104] developed a series of chirality indices using an approach called MARCH-INSIDE (Chiral Markovian chemicals "in silico" design). This approach is based on a one-step electron transition stochastic matrix $^1\Pi$ of order n, (the number of atoms in a molecule); the elements of $^1\Pi$ are $^1p_{ij}$ (transition probabilities derived from electronegativities of atoms).

$$^np_{ij} = \frac{\chi_j}{\sum_{k=1}^{\delta+1} \chi_k} \tag{12}$$

where χ_j is the electronegativity of atom j bonded to atom i. The MARCH-INSIDE descriptors are then

$$^{SR}\pi_k(G) = \sum_{i=1}^{g} {}^kp_{ij} = Tr\left[\left(^1\pi\right)^k\right] \tag{13}$$

The graph G may be that of a whole molecule (molecular graph) or any fragment (subgraph). As Eq. 12 does not contain any parameter to encode chirality a dummy variable ω_k was incorporated into Eq. 12. The chirality variable ω_k is defined as

$$^1p(\omega_{ij}) = \frac{\chi_j e^{\omega_j}}{\sum_{k=1}^{\delta+1} \chi_k e^{\omega_k}} \tag{14}$$

The chirality variable ω_k takes the value 1 for R-configuration, -1 for S-configuration and becomes 0 for achiral molecules. Equation (13) is modified to Eq. 15 for calculating the chiral descriptor

$$^{SR}\pi_k(G, \omega) = \sum_{i=1}^{g} {}^kp_{ij}(\omega) = Tr\left[\{^1\pi(\omega)\}^k\right] \tag{15}$$

Fig. 9 Structure of perindopril. The molecule has 5 chiral centers hence, $2^5 = 32$ stereoisomers

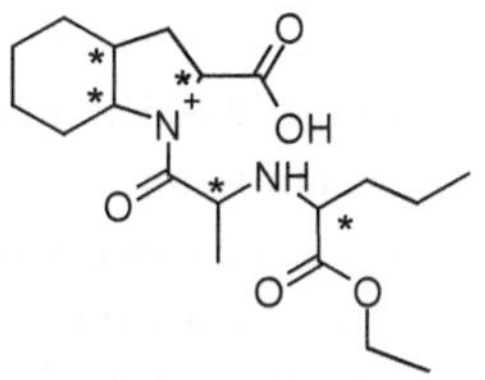

(11)

Perindopril

If the molecule is achiral Eq. 14 will automatically become Eq. 12 and Eq. 15 will be reduced to Eq. 13. Application of MARCH-INSIDE descriptors in QSAR modeling was performed the classification of ACE activity of thirty-two perindopril (**11**) stereoisomers and the fourteen (7 pairs of enantiomers) dopamine σ-receptor antagonist activity of (3HPP). The MARCH-INSODE approach was extended for pseudographs considering either atom (vertex) adjacency [13] or bond (edge) adjacency [105] (Fig. 9).

3.5   Relative Chirality Indices (RCI)

Relative chirality indices (RCI) were introduced by [106] in 2007. Though this approach uses vertex degrees to compute the new chirality indices it does not include any chiral modification. In order to calculate RCI for a chiral molecule the following steps are followed:

- CIP rules are applied to assign priorities for the four ligands that are attached to the chiral center(s) and then they are ranked in the decreasing order as *A, B, C* and *D* (*A* has the highest priority).
- Atoms in each of the four ligands (sometimes it may be a single atom) are then assigned valence delta-values of atoms (δ^v) according to the method of Hall and Kier [97].
- When a ligand contains more than a single atom, δ^v for the ligand (*A, B, C* or *D*) is calculated taking into account the through bond distance i.e., the relative proximities of the atoms to the chiral center. A decreasing importance with increasing topological distance (through bond) was assigned while calculating the contribution of atoms other than hydrogen in a group. The group delta value for any group ($\delta^v{}_i$) attached to a chiral carbon is calculated as:

$$\delta_i^v = \delta_{n1}^v + \frac{\delta_{n2}^v}{2} + \frac{\delta_{n3}^v}{4} + \frac{\delta_{n4}^v}{8} + \dots \tag{16}$$

where n_1 is the atom directly attached to the chiral center, n_2 is seconds atom in the ligand from the chiral center and so on. The weighting the atoms according to their distance to the central atom is significant because the farther the atom is from the chiral center, the lower is its contribution and this was shown by [107] that optical activity of analogues reaches a constant value with increase in carbon atoms.

The transformation of information encoded in the 3-D structure of a chiral molecule and the calculation of relative chirality indices is outlined in Sect. 3.8 (Fig. 11). A chiral molecule with one chiral center is considered. The relative sizes of the R_1, R_2 and R_3 indicate the order of priority of the three substituents (assigned as per CIP rule) attached to the chiral carbon such that $R_1 > R_2 > R_3$ and the fourth substituent of least priority is taken as hydrogen. The least priority group is placed away from the viewer and the configuration is assigned as R or S based on the clockwise or anti-clockwise direction of arrangement of the three substituents. The 3D structures may thus be represented by two directed arcs where the three groups R_1, R_2 and R_3 represented by A, B and C, respectively (Fig. 10).

In order to differentiate the two directed arcs, and consequently the enantiomers two formulas are proposed. The new chirality indices thus generated are called the relative chirality indices $^R RCI$ and $^S RCI$. The formulas to calculate $^R RCI$ and $^S RCI$ are given below:

$$^R RCI = \delta_a + \{\delta_a + (\delta_a \times \delta_b)\} + \{\delta_a + (\delta_a \times \delta_b) + (\delta_a \times \delta_b \times \delta_c)\} + (\delta_a \times \delta_b \times \delta_c \times \delta_d)$$
(17)

$$^S RCI = \delta_a + \{\delta_a + (\delta_a \times \delta_c)\} + \{\delta_a + (\delta_a \times \delta_c) + (\delta_a \times \delta_b \times \delta_c)\} + (\delta_a \times \delta_b \times \delta_c \times \delta_d)$$
(18)

where δ_a, δ_b, δ_c and δ_d are the weights assigned to the four substituents attached the chiral center and the priority order (assigned as per CIP rules) of the four substituents are A > B > C > D.

Information regarding the lowest priority (D) is also included in the calculation of RCI and is taken care of by the term $\delta_a \times \delta_b \times \delta_c \times \delta_d$. If the fourth group D happens to be hydrogen, this term reduces to zero. If more than one chiral center is present RCI is calculated for chiral center and root mean square of these indices for chiral centers gives the RCI for the diastereomers.

If more than one chiral center is present then $^R RCI$ and $^S RCI$ are calculated for each chiral center. RCI for the diastereomers are calculated by taking root-mean square of the RCI for all centers.

$$RCI = \sqrt{\frac{1}{N} \sum_{i=1}^{N} (RCI_i)^2}$$
(19)

The weight of a vertex δ may be calculated based on valence connectivity of the atoms, formula mass of the atoms, electro-topological states etc., to compute RCI on different

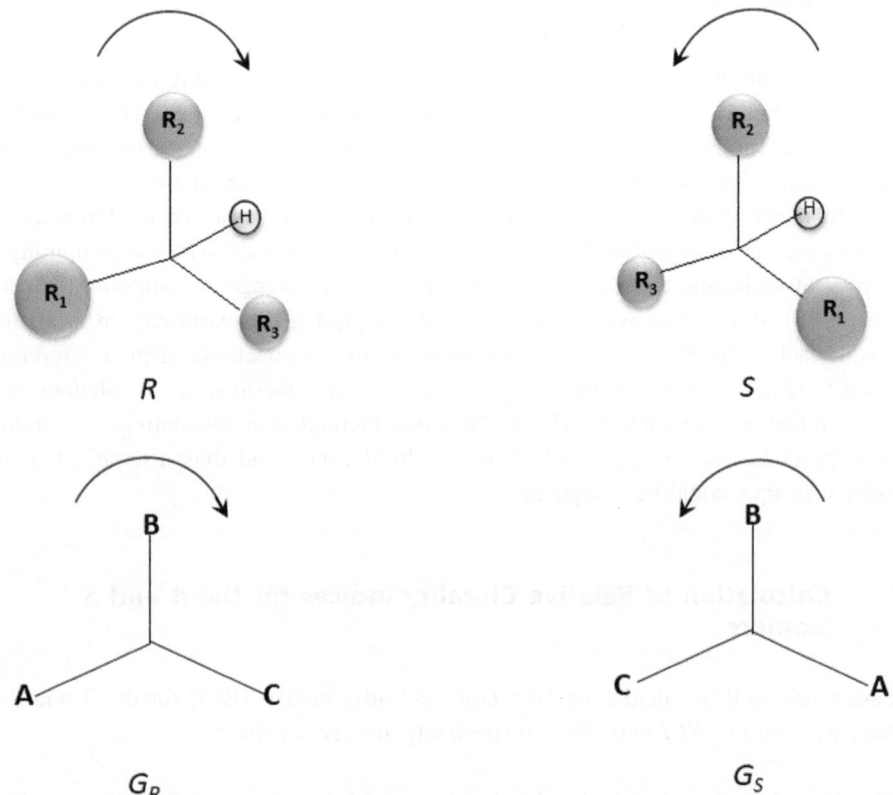

Fig. 10 3D structure to 2D directed arcs

scales. Illustrations for calculating RCI are given in Sect. 3.8. ^{R}RCI and ^{S}RCI could distinguish the enantiomers and diastereomers. *Meso* isomers have the same *RCI* values for example the *meso* tartaric acids were found to have the same values irrespective of the algorithm used to assign vertex weights. The RCI were used to order the repellency of the four diastereomers of AI3-37220. In the case of the 7 pairs of 3HPPs, reasonably good QSAR model were obtained for both the σ receptor and dopamine D2 receptor affinities. This data set of 3HPPs is used by several other authors to estimate the performance the chirality indices proposed. This enabled the comparison of the performance of RCI in modelling the σ receptor and dopamine D2 receptor affinities with other chirality descriptors [38].

3.6   Modified Relative Chirality Indices

In the calculation of RCI discussed in the previous section decreasing importance to the atoms in a group as one moves away from the chiral center is applied. This is modified and in the new approach equal importance is given to all the vertices (atoms) because the atoms away from the chiral carbon are more flexible and are likely to play a role in binding to an active site in a protein. Moreover, it becomes difficult to extend to other type of indices such as information theoretic indices and available software for computing the large pool of molecular descriptors could not be taken advantage in computing the chiral indices (RCI). In order to avoid this, in the new approach the chiral carbon is included as part of each of the four ligands so that the molecular connectivity of the atom directly attached to chiral carbon remains unaltered, and this also facilitate us to calculate many chirality indices for a molecule. The modification facilitated in calculating large number of descriptors for the groups attached to the chiral center and thus a large set of new chirality measures could be computed.

3.7   Calculation of Relative Chirality Indices for the R and S Isomers

The steps followed in calculating the relative chirality indices (RCI) for the R and the S isomers denoted as ^{R}RCI and ^{S}RCI, respectively are given below:

1. The four groups attached to the chiral carbon are assigned priority following the Cahn, Ingold and Prelog (CIP) system as a, b, and d, where is the highest priority and d is the lowest priority group.
2. SMILES notation [108], Weininger et al. [109] is written for each group of the four groups and in order to maintain the connectivity of the atom (vertex) connected to the chiral carbon chiral carbon is also included to the groups. This is illustrated in the following examples:
3. Various molecular descriptors are calculated for each of the four substituents (a, b, c, and d) using the SMILES notations as the input. The computer program *INDCAL* [110], *POLLY* [77, 78], *TRIPLET* (the computer program *TRIPLET* computes the triplet indices proposed by [111, 112] were used to compute the molecular descriptors for each of the four substituents.
4. The descriptors thus calculated for each substituent a, b, c and d are taken as the weight of that substituent namely, δ_a, δ_b, δ_c, and δ_d.
5. Relative Chirality Indices (RCI) for are then calculated for the R and S isomers using Eqs. 17 and 18 for the R and the S isomers, respectively

6. Based on the type of molecular descriptor assigned to the substituents the RCI are denoted as $^R RCI_{V0}$, $^R RCI_{B1}$, $^S RCI_{SUM\text{-}R}$, $^R RCI_{OPM}$ etc., where the subscript indicates the type of molecular descriptor used in assigning the weights (δ) to the four substituents attached to the chiral carbon.

7. As mentioned above, if more than one chiral center is present then $^R RCI$ and $^S RCI$ are calculated for each chiral center. RCI for the diastereomers are calculated by taking root-mean square of the RCI for all centers using Eq. 19.

For instance, if a molecule has two chiral centers and they are the 2nd and the 3rd carbon atoms of the molecule then the molecule will have four diastereomers namely, (2R, 3R), (2S, 3R), (2R, 3S) and (2S, 3S). RCI for each of these diastereomers $^{2R3R} RCI$, $^{2S3R} RCI$, $^{2R3S} RCI$ and $^{2S3S} RCI$ may be calculated as below

$$^{2R3R} RCI \sqrt{\frac{\left(^{2R} RCI^2\right) + \left(^{3R} RCI^2\right)}{2}} \tag{20}$$

$$^{2S3R} RCI \sqrt{\frac{\left(^{2S} RCI^2\right) + \left(^{3R} RCI^2\right)}{2}} \tag{21}$$

Similarly, *RCI* is calculated for the other two stereoisomers.

3.8　　Illustrations for the Calculation of Relative Chirality Indices

(1) Using simple connectivity index $^1\chi$ (S$_1$)

$$^R RCI_{S1} = (1) + \{(1) + (1 \times 1.732)\} + \{(1) + (1 \times 1.732) + (1 \times 1.7321 \times 4.326)\}$$
$$+ 0 = 13.956$$

$$^S RCI_{S1} = (1) + \{(1) + (1 \times 4.326)\} + \{(1) + (1 \times 4.326) + (1 \times 1.7321 \times 4.326)\}$$
$$+ 0 = 19.144$$

(2) Using Balaban Index J

$$^R RCI_J = (1) + \{(1) + (1 \times 3.098)\} + \{(1) + (1 \times 3.098) + (1 \times 3.098 \times 2.907)\}$$
$$+ 0 = 18.204$$

$$^S RCI_J = (1) + \{(1) + (1 \times 2.907)\} + \{(1) + (1 \times 2.907) + (1 \times 3.098 \times 2.907)\}$$
$$+ 0 = 17.822$$

(3) Using Shape Index $^1\kappa$

$$^RRCI_{1\kappa} = (1) + \{(1) + (1 \times 2)\} + \{(1) + (1 \times 2) + (1 \times 2 \times 3.556)\}$$
$$+ 0 = 13.111$$

$$^SRCI_{1\kappa} = (1) + \{(1) + (1 \times 3.556)\} + \{(1) + (1 \times 3.556) + (1 \times 2 \times 3.556)\}$$
$$+ 0 = 17.222$$

The modification in RCI enables the computation of a large pool descriptors based on simple connectivity, valence connectivity, information theory, triplet state indices, shape indices etc. As different RCIs encode different information about a chiral molecule it become easy to handle data sets containing structurally dissimilar molecules. The CIP rule assigns *R*- and *S*-configurations to a chiral center based on a qualitative dichotomous approach. But RCI approach gives a quantitative scale to a set of chiral molecules based on their structural characteristics. This is similar to the computation of numerical graph invariants or topological indices of molecular graphs whereby qualitative concepts like branching, complexity are transformed into quantitative scales [113] (Fig. 11 and Table 2).

Fig. 11 Applying the sequence rule to tyrosine

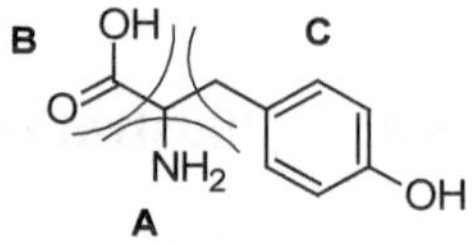

(12)
Tyrosine

Table 2 Substituents and their weights (δ_i) calculated on different principles of connectivity

Group	Substituent	SMILES including chiral C	Values some descriptors calculated (δ)		
			$^1\chi\ (S_1)$	J	$^1\kappa$
A	–NH2	CN	1.000	1.000	1.000
B	–COOH	CC(=O)O	1.732	3.098	2.000
C	–CH2C6H4–OH(*p*)	CCc1ccc(O)cc1	4.326	2.191	3.556
D	–H	–	0	0	0

^{R} *RCI* and ^{S} *RCI* can be calculated using Eq. 17 and Eq. 18, respectively and the weights (δ) computed for each substituent

4 Cartesian Coordinate of Heteroatom-Based Descriptor

Usually, chemicals graphs are converted into adjacency and distance matrices and hence, molecular chirality descriptors are based on these matrices. Diudea and Ursu developed a descriptor based on the internal cartesian coordinates of the molecule (Å). The chirality of an ordered quadruple of atoms numbered 1, 2, 3, 4 is measured in terms of their (x, y, z) Cartesian coordinates, adopting some geometrical constraints, by the sign of the following determinant [114, 115]:

$$C_{1234} = \text{sign} * \det \begin{matrix} x_1 & x_2 & x_3 \\ y_1 & y_2 & y_3 \\ z_1 & z_2 & z_3 \end{matrix}$$

$$C_{1234} = sign \times [A]$$

where $A = \begin{vmatrix} x_1 & x_2 & x_3 \\ y_1 & y_2 & y_3 \\ z_1 & z_2 & z_3 \end{vmatrix}$

The value of the determinant is discriminant for each molecule. The sign of the resulting determinant should be sensitive to chirality. Furthermore, most bioactive molecules are chiral. Cluj centrality descriptor [115, 116] allows the finding of the graph center (e.g., the vertex having the largest C_i value) and provides an ordering of graph vertices according to their centrality best describes bioactivity (it focuses on heteroatoms where the active chemical center of the molecule is chiral. Hence, Cluj centrality index was added when organic compounds are chiral.

Furthermore, a point symmetrical to another point is a point. Point (A) symmetrical to point (O) is point (A′), located along the extension of the line AO so that AO = OA′. The symmetric of point A concerning line (d) is the point (A′) located on the extension of the perpendicular from (A) to (d) such that AO = OA′. The symmetry of a surface concerning an axial axis is another surface obtained by the symmetrical union of the given surface points. The symmetry of the surface ABC concerning axis (d) is the surface A′B′C′ obtained by joining the symmetries of the points A′B′C′, symmetrical to the points A, B, and C. The critical property of the symmetry of the surface is the equality of the two surfaces: given and its symmetry, Area ABC = Area A′B′C′. Analytical geometry provides the coordinates of the points of the given and symmetrical surfaces. The distances between different points of the surfaces can be determined. Thus, the areas of the two surfaces can be established by observing their equality. It is observed that a surface is represented by a number that is the area of that surface. The theory of algebraic structures developed in the contemporary period has become essential for topological and differential structures. An algebraic structure is a set on which some operations have been defined: addition, multiplication, intersection, and reunion. In topology, the connection of a structure is essential. A function is bijective if its derivative is different from zero. This is valid only in the definition domain of the function and if it is convex. A matrix represents a linear application. It is valid only for assembly; it multiplies the scope of the product. Of all the square matrices, having the number of rows (m) equal to that of the columns (n), m = n has the property of symmetry and complementarity. If a matrix (A) is equal to the transposed matrix (A–): A = A–, then the matrix (A) is symmetric. The complement of a matrix (A) is denoted by Ac or (A). A multimer (A′) is the complement of many sets (A) if the elements of the set (A′) are not contained in set A. A determinant of the same order can be associated with the matrix. An additional minor M_{ij} of an element A_{ij} is the determinant obtained by deleting the line (i) and the column (J). The algebraic complement of an element (A_{ij}) is:

$$A_{ij} = (-1)^{i+j} M_{ij}. \tag{22}$$

The determinant of matrix A is the sum of the products of all the elements of a row or column and their algebraic complements:

$$\det A = \sum_{i=1}^{n} a_{ij} A_{ij} \sum_{j=1}^{n} a_{ij} A_{ij} \tag{23}$$

$$\Delta = \sum a_{ij}\, M_{ij} \tag{24}$$

$$= \sum \sum a_{ij(-1)^{i+j}} M_{ij} \tag{25}$$

Thus, a chirality index can be computed using a square matrix composed of the atom's internal Cartesian coordinates computed for the heteroatom and the neighbour atoms.

By using this algorithm, the c chirality index was computed for the 7 pairs of enantiomers of 3-(3-hyrdoxyphenyl)piperidines. The structures of the molecules and the values of chirality index c computed for the 14 molecules (7 pairs of R and S isomers) are given in Table 3.

The correlation between the chirality descriptor obtained from the Cartesian coordinates and the biological activities of the seven pairs of enantiomers brings out an important aspect. Correlation coefficient r with σ receptor affinities is −0.7780 whereas for the D2 receptor affinities the correlation coefficient r = 0.1647. That is, with one biological activity of a set of molecules the correlation is reasonably good whereas for the other biological property of the same set of molecules the correlation is poor. If one investigates the relation between the two biological properties (σ receptor affinity vs. D2 receptor affinity), they are mutually unrelated indicated by the poor correlation between them (r = 0.1947). Therefore, prediction or modeling mutually unrelated biological or any physicochemical property with one molecular descriptor is not possible. Under such situations a diverse

Table 3 Structures of 3HPPs, the new chiral descriptors and the dopamine (DA) D2 and opiate σ receptor affinities

No	R =	Config	New Chirality descriptor	pIC$_{50}$ of μM	
				σ receptor	D2 receptor
1	H	R	−1.6950	−0.66	0.36
2		S	1.4912	−1.19	−0.24
3	−CH$_3$	R	−1.9309	0.43	−0.61
4		S	1.6533	−0.28	−0.29
5	−CH$_2$CH$_3$	R	−3.4957	0.95	−0.54
6		S	0.7926	−0.01	−0.12
7	−CH$_2$CH$_2$CH$_3$	R	−4.8122	1.52	−0.23
8		S	−1.4444	0.81	0.38
9	−CH(CH$_3$)$_2$	R	−4.4941	0.61	−1.08
10		S	0.3257	0.68	0.45
11	−CH$_2$ CH$_2$CH$_2$CH$_3$	R	−5.8005	2.05	−0.43
12		S	−1.2521	1.51	0.56
13	−CH$_2$CH$_2$-C$_6$H$_5$	R	−7.6844	2.10	−0.09
14		S	−3.4464	1.80	1.36

pool of descriptors as in the case of RCI has better chances in the development of quality prediction models.

5 Variation of Chemical Space Explored by Riemann Surfaces

Before going into the details of tracing the Reimann surfaces for a molecule, a simplified definition is presented here. If the function is defined as multi-valued with the domain on the real axis, then there is an ambiguity in its solution. To avoid this ambiguity, the idea of Riemann surface theory comes into picture. The idea is to replace the domain of a multi-valued function and its graph into a complex plane or z-plane and study the function on the w-plane, known as the Riemann surface.

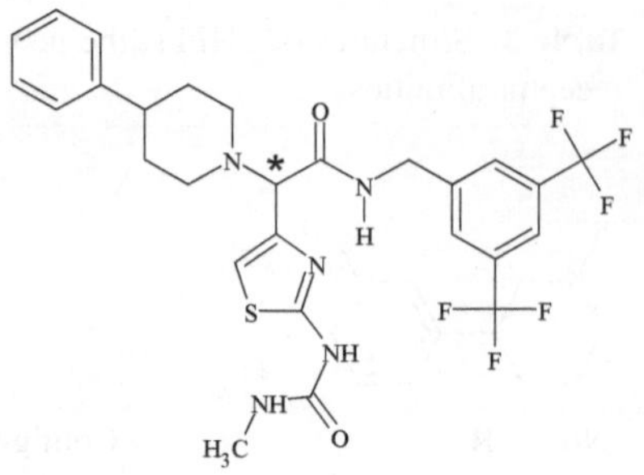

Fig. 12 Example of a chiral
molecule with heteroatoms
(chiral carbon is marked)

(13)

The first step in computing the Reimann surfaces is to compute the Cartesian coordinates for the heteroatoms as explained in the previous section. This is illustrated with the chiral molecule **13** that has not only one chiral center but also the heteroatoms N, S and O in its molecule (Figs. 12, 13 and Table 4).

The descriptors obtained from the Cartesian coordinates (as in the case of 3HPPs) for the R- and S-isomers are 16.900 and 237.09, respectively. The conversion of Cartesian coordinates into Reimann surfaces is explained further here.

5.1   Reimann Surfaces

The need to determine the surfaces requires a matrix mathematical tool. It is known that the solutions of a system of linear equations depend on the coefficients of the system. Therefore, a matrix is obtained by making a rectangular table of these coefficients arranged by lines and columns.

$$A = \begin{pmatrix} a_{11}a_{12}\ldots a_{1n} \\ \cdots\cdots\cdots\cdots\cdots\cdots \\ a_{m1}a_{m2}\ldots a_{mn} \end{pmatrix} \tag{26}$$

This array with (m) lines and (n) columns is an (A) matrix or $m \times n$ matrix. If $m = n$, the matrix is called a square. The set of $m \times n$ matrices form a vector space of dimensions (m × n). The null vector of this space is the null matrix with all null coefficients. In addition, multiplication and division operations can be applied to the matrices with computability and associativity properties.

If a square matrix has zero coefficients except for those on the main diagonal where they are equal to unity, it is called a unity matrix or an identical matrix; it is of the form:

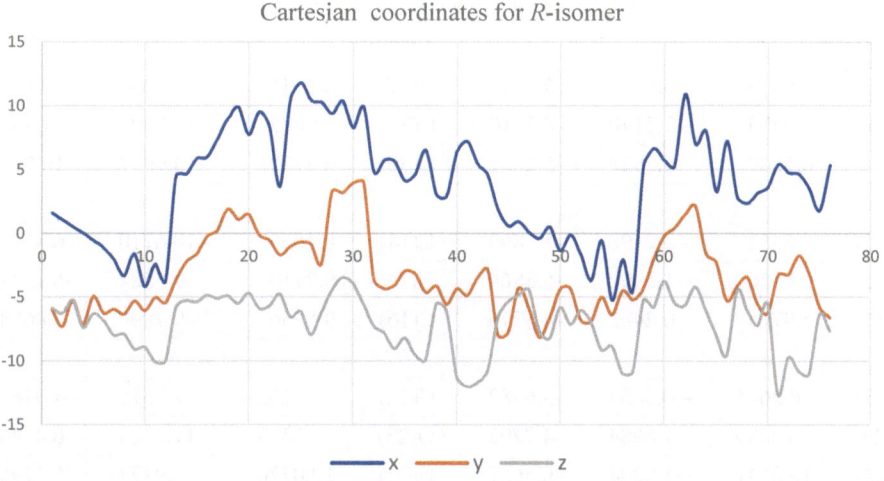

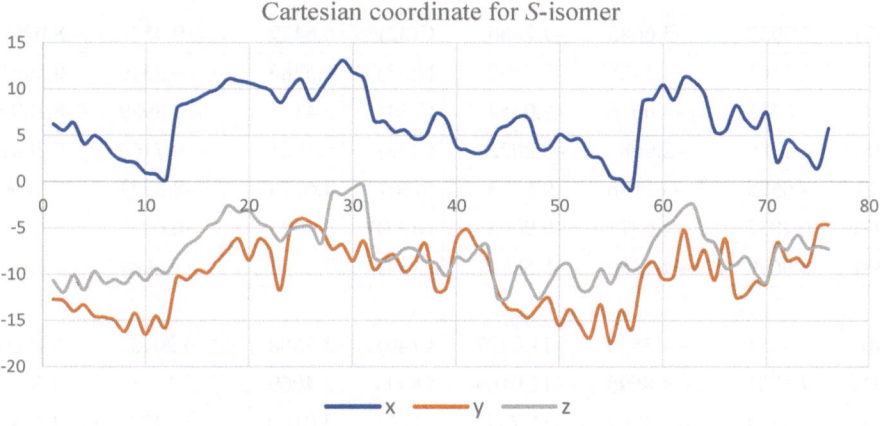

Fig. 13 Cartesian coordinates of the R- and S-isomers compound **13**

$$I = \begin{pmatrix} 10\ldots0 \\ 01\ldots0 \\ 00\ldots1 \end{pmatrix} \tag{27}$$

Each square matrix can be associated with a real number called a determinant. Thus, the matrix A has the corresponding determinant:

$$\det A = \begin{vmatrix} a_{11}\ldots a_{1n} \\ \ldots\ldots\ldots\ldots \\ a_{m1}\ldots a_{mn} \end{vmatrix} \tag{28}$$

Table 4 Cartesian coordinates of the atoms in the R- and S-isomers of compound **13**

R isomer				S isomer			
N(1)	1.5829	−6.0259	−5.9049	N(1)	6.2345	−12.7050	−10.6882
C(2)	1.0589	−7.3140	−6.2710	C(2)	5.5675	−12.8891	−11.9488
C(3)	0.5209	−5.2828	−5.2822	C(3)	6.3720	−13.9977	−10.0735
...				...			
C(14)	4.6351	−2.2498	−5.2890	C(14)	8.4645	−10.6110	−6.8677
N(15)	5.8356	−1.5992	−5.3864	N(15)	8.9710	−9.5584	−6.1538
C(16)	5.9715	−0.2482	−4.8776	C(16)	9.5956	−9.7896	−4.8658
...				...			
C(22)	8.2641	−0.5654	−5.6982	C(22)	9.8703	−7.3519	−4.9765
O(23)	3.6889	−1.6984	−4.7792	O(23)	8.5379	−11.7324	−6.4246
C(24)	10.5141	−1.0294	−6.5657	C(24)	10.0720	−4.9177	−5.2295
...				...			
C(32)	4.7937	−3.6685	−7.2850	C(32)	6.6439	−9.4532	−8.0369
N(33)	5.7161	−4.3406	−7.8190	N(33)	6.5285	−8.3310	−8.5981
C(34)	5.6215	−4.0676	−9.0454	C(34)	5.4175	−7.8989	−8.1899
C(35)	4.1186	−2.9681	−8.2022	C(35)	5.5821	−9.7160	−7.2681
S(36)	4.6585	−3.1986	−9.5379	S(36)	4.6179	−8.6221	−7.3166
N(37)	6.4832	−4.6378	−9.9063	N(37)	4.9923	−6.6959	−8.6151
C(38)	3.0699	−4.1450	−5.5881	C(38)	7.3393	−11.6961	−8.7879
...				...			
C(40)	6.4011	−4.3573	−11.2437	C(40)	3.7888	−6.2043	−8.1860
O(41)	7.1751	−4.8695	−12.0169	O(41)	3.4069	−5.1238	−8.5679
N(42)	5.4418	−3.4964	−11.7049	N(42)	3.0180	−6.9373	−7.3242
C(43)	4.5127	−2.8816	−10.7768	C(43)	3.4764	−8.2343	−6.8657
...				...			

Cramer's rule solves a linear system of (quadratic) linear equations and (quadratic) unknowns Given a linear system of two equations with two unknowns, the following are true:

$$\left\{\begin{array}{l} a_1x + b_1y = c_1 \\ a_2x + b_2y = c_2 \end{array}\right\}$$

If $\Delta \begin{vmatrix} a_1b_1 \\ a_2b_2 \end{vmatrix} \neq 0$ and applying Cramer's rue, the following system of equations are obtained:

$x = \frac{\Delta X}{\Delta}, y = \frac{\Delta y}{\Delta}$, where $\Delta x = \begin{vmatrix} c_1b_1 \\ c_2b_2 \end{vmatrix}$, $\Delta y = \begin{vmatrix} a_1c_1 \\ a_2c_2 \end{vmatrix}$, $\Delta = \begin{vmatrix} a_1b_1 \\ a_2b_2 \end{vmatrix}$

In order to associate a matrix to a linear transformation, a base $(x_1...,x_n)$ is chosen. The images of this base are:

$$A(x_j) = \sum_{i=1}^{m} a_{ij} for j = 1 \ldots n$$

Retrieving the square matrix of the transformation. $A \rightarrow A = \begin{pmatrix} a_{11}....a_{1n} \\ .... \\ a_{m1}....a_{mn} \end{pmatrix}$.

The existence of a base for which the matrix represents a transformation is closely related to the eigenvalue's theorem.

A number (λ) is an eigenvalue (or the characteristic value) of a linear transformation (A) if the vector $x \neq 0$ and $A(x) = \lambda x$.

The vector (x) is called an eigenvector of the A transformation related to (λ).

By rewriting the equation, $A(x) = \lambda x$ like $(A-\lambda x) = 0$, λ is an eigenvalue of the A operator if and only if the operator $(a-\lambda j)$ is singular.

The eigenvalues of the $(A-\lambda j)$ operator are the solutions of its corresponding determinant. Symmetric transformation's eigenvalues are real, existing a real eigenvectors base.

In conclusion, if a base of a vectorial space is chosen then the equation $(A-\lambda j)(x) = 0$ can be written like an equation system of the (x) coordinates x_1, x_n:

$$(a_{11} - \lambda)x_1 + a_{12}x_2 + \ldots + a_{1n}x_n = 0$$
$$A_{21}x_1 + (a_{22} - \lambda) x_2 + \ldots + a_{2n}x_n = 0$$
$$......$$
$$A_{n1}x_1 + a_{n2}x_2 + \ldots + (a_{nn} - \lambda) x_n = 0$$

The coefficient matrix is the matrix A-λI, which represents the transformation A-λJ. Only the nonnull vectors can be eigenvalues. Determining the nonnull solution of the homogenous system can be obtained, as shown before.

Overall, the determinant $\det(A-\lambda J) = a_0\text{-}a_1\lambda + \ldots + a_n\lambda_n$ defines the characteristic polynomial.

The coordinates $x_1, x_2,\ldots,x_n$ of (x) are the solutions of the above homogenous system, where λ is a solution of the characteristic polynomial.

In a rectangular axis system in the xOy plane, $y = f(x)$ has the surface area between Ox and f(x). This area represents a natural number ($N \geq 0$). Determination of this surface is performed by using the formula. $\int f(x)\ dx$. If only the determination of a surface contained between two points on the Ox axis (and part of (f) x) is needed, the formula of Newton Leibniz can be applied:

$$A = \int_a^b f(x)dx = F(a) - F(b) + C, \text{ or,}$$

$$A\prime = \int_a^b f(x)dx = F(x) \mid_a^b \text{ while } F(x) \mid_a^b = F(b) - F(a)$$

The approximation of a function area on an interval is performed by dividing its area in rectangular equals along Ox in length:

$x_2\text{-}x_1 = x_3\text{-}x_2 = x_4\text{-}x_3,$ while the heights of the rectangles are determined by the contact points with the function (A, B, C,), y_2, y_3, y_4.

The function's relative surface is determined by summing the surfaces of the three rectangles. These area sums are called Riemann sums.

$$A = \sum_{i=1}^n f(xi)\Delta x,$$

where values of $f(x_i)$ are y_2, y_3, y_4 and Δx is $x_2\text{-}x_1, x_3\text{-}x_2, x_4\text{-}x_3,$ as shown above.

Riemann surfaces show variations in space and are complex regarding their geometry. The simplest Riemann surface is the sphere of Riemann numbers. Other examples are the hyperbolics, the tor, and Mobius surfaces.

A Riemann surface (in topology) has a countable base of a complex structure, is orientable, keeps its orientation (Mobiuous band), and can have a conformal-conformational structure (same angles). A curb surface can be elliptical, plane, or hyperbolic.

In Riemann's topology, the scalar curve is discussed. Each point corresponds to a real number. The scalar curvature of a sphere of radius r is equal to $\frac{2}{r^2}$.

The transverse curvature of a sphere of radius r is $k = \frac{1}{r^2}$ and determines a scalar curvature $S = \frac{n(n-1)}{r^2}$, where (n) is the n-dimensional space.

In hyperbolic, n-dimensional space, the transverse curvature is $K=\frac{1}{r^2}$ and determines a scalar curvature of $S = -\frac{n(n-1)}{r^2}$.

Overall, the distinct Riemann surfaces are obtained for each enantiomer regardless of their external coordinates' values. Each Riemann surfaces is distinct for each heteroatom depending on the stereoisomer namely, *S*- or *R*- enantiomer.

5.2 Reimann Surfaces from Cartesian Coordinates of Enantiomers

Using the cartesian coordinates, the following square matrices were obtained concerning each heteroatom for the *R*-isomer of **13**:

1.5829	−6.0259	−5.9049	4.6351	−2.2498	−5.2890	8.2641	−0.5654	−5.6982	4.7937	−3.6685	−7.2850
1.0589	−7.3140	−6.2710	5.8356	−1.5992	−5.3864	3.6889	−1.6984	−4.7792	5.7161	−4.3406	−7.8190
0.5209	−5.2828	−5.2822	5.9715	−0.2482	−4.8776	10.5141	−1.0294	−6.5657	5.6215	−4.0676	−9.0454

4.1186	−2.9681	−8.2022	4.6585	−3.1986	−9.5379	6.4011	−4.3573	−11.2437	7.1751	−4.8695	−12.0169
4.6585	−3.1986	−9.5379	6.4832	−4.6378	−9.9063	7.1751	−4.8695	−12.0169	5.4418	−3.4964	−11.7049
6.4832	−4.6378	−9.9063	3.0699	−4.1450	−5.5881	5.4418	−3.4964	−11.7049	4.5127	−2.8816	−10.7768

The square matrices for the *S*-isomer of **13** are:

6.2345	−12.7050	−10.6882	8.4645	−10.6110	−6.8677	9.8703	−7.3519	−4.9765	6.6439	−9.4532	−8.0369
5.5675	−12.8891	−11.9488	8.9710	−9.5584	−6.1538	8.5379	−11.7324	−6.4246	6.5285	−8.3310	−8.5981
6.3720	−13.9977	−10.0735	9.5956	−9.7896	−4.8658	10.0720	−4.9177	−5.2295	5.4175	−7.8989	−8.1899

5.5821	−9.7160	−7.2681	4.6179	−8.6221	−7.3166	3.7888	−6.2043	−8.1860	3.4069	−5.1238	−8.5679
4.6179	−8.6221	−7.3166	4.9923	−6.9590	−8.6151	3.4069	−5.1238	−8.5679	3.0180	−6.9373	−7.3242
4.9923	−6.6959	−8.6151	7.3393	−11.6961	−8.7879	3.0180	−6.9373	−7.3242	3.4764	−8.2343	−6.8657

Characteristic polynomials were computed for the square matrices and the following equations were obtained for each square matrix for *R*-isomer. By converting λ (characteristic polynomial) to x, the following equations were obtained (Table 5).

Similarly, the characteristic polynomials for the *S*-isomer of **13** are given below in Table 6.

By using these equations, Riemann surfaces were computed for each heteroatom matrix by computing and plotting the square root of each equation (for the two enantiomers) (Table 7).

The global descriptor computed for each molecule by using the square matrix determinant sum is as follows:

For *R*-isomer

$$D = (5.2288) + (-4.5622) + (-13.9013) + (-1.05642) + (1.9995) + (31.3528)$$
$$+ (-0.9818) + (-1.1793)$$

Table 5 Characteristic polynomial equation and (Δ) the polynomial disicrminant for R-isomer of 13

	Δ	Characteristic polynomial
1	35,856	$-x^3 - 11.0133x^2 + 4.97626506x + 5.22885$
2	$-33,828.5$	$-x^3 - 1.8417x^2 - 21.1549341x - 4.56221$
3	-5217.64	$-x^3 + 0.0667237x - 13.90130$ $(-x^3 + 5.3290 \times 10^{-15}x^2 - 0.0667x - 13.9013)$
4	21,477.2 -420.233	$-x^3 - 8.3112x^2 + 12.29491686x + 0.87572$ $(-x^3 - 8.5923x^2 - 5.21157x - 1.05642)$
5	5558.82	$-x^3 - 8.9863x^2 - 0.48097431x + 1.99955$
6	48,579.4	$-x^3 - 5.56746x^2 + 12.76491x + 31.35295$
7	-3745.36	$-x^3 - 10.1733x^2 - 1.33675944x - 0.9818$
8	33,751.8	$-x^3 - 7.0981x^2 + 17.73296391x - 1.17935$

Table 6 Characteristic polynomial equation and (Δ) the polynomial disicrminant for S-isomer of 13

	Δ	Characteristic polynomial
1	32,647.3	$-x^3 - 16.7281x^2 + 41.73740071x - 23.35831$
2	$-91,027$	$-x3 - 5.9597x^2 - 25.26336446x + 20.38310$
3	272,388	$-x^3 - 7.0916x^2 + 24.76561618x + 62.10049$
4	$-53,435$	$-x^3 - 9.877x^2 + 4.19356075x - 11.30411$
5	87,219.1	$-x^3 - 11.6551x^2 - 10.22130968x + 21.40881$
6	331,202	$-x^3 - 11.129x^2 + 15.58285231x + 37.52168$
7	22,276.8	$-x^3 - 8.6592x^2 + 23.23056144x - 10.51012$
8	121,208	$-x^3 - 10.3961x^2 + 14.45660419x - 12.61621$

$$D = 16.9001$$

For S-isomer

$$D = (-23.3583) + (20.3831) + (62.1004) + (-11.3041) + (21.4088) + (37.52168)$$
$$+ (-10.5101) + (-12.6162)$$

$$D = 83.62533$$

For the R-isomer, the overall square matrix product including the eight heteroatoms is:

Table 7 Riemann surfaces with of real (blue) and complex (red) roots

1	N		1	N	
2	N		2	N	
3	O		3	O	
4	N		4	N	
5	S		5	S	
6	N		6	N	
7	O		7	O	
8	N		8	N	

$$-164424\ 78430.3\ 211534$$
$$-187509\ 89439.0\ 241223$$
$$-185339\ 88394.6\ 238397$$

The characteristic polynomial computed is:

$$-\lambda^3 + 163412\lambda^2 - 6.95937 \times 10^6\lambda - 444884$$

For the S-isomer the overall square matrix product including the eight heteroatoms is:

$$-5.33209 \times 10^7\ 1.37357 \times 10^8\ 1.09377 \times 10^8$$
$$-6.06091 \times 10^7\ 1.56132 \times 10^8\ 1.24327 \times 10^8$$
$$-5.49779 \times 10^7\ 1.41626 \times 10^8\ 1.12776 \times 10^8$$

where the computed characteristic polynomial is (Table 8)

$$-\lambda^3 + 2.15587 \times 10^8\lambda^2 + 7.9302 \times 10^9\lambda - 6.9365 \times 10^{12}$$

6   Chiral Distinction Under an Applied Field

In this section, chiral distinction of enantiomers under applied fields is discussed. Though these methods are not directly related to the numerical characterization of chiral molecules, the theme of this review, these recent developments are presented here to highlight the emerging methods of enantio-detection and chiral analysis. It is well known that only polarized light can differentiate enantiomers and similarly a chiral receptor can only recognize and differentiate chiral isomers. Same is true for energies of enantiomers calculated in vacuum using quantum chemical calculations. Enantiomers have the same energy irrespective of the basis set used for molecular optimization. However, when the energies of the enantiomers are computed in an applied electric field the enantiomers differ in their energies owing to the change in electron distribution and frontier orbitals. Chen et al. [117] studied the effect of chiral molecules in an external electric field. In their study they carried of quantum chemical calculations for 74 pairs of enantiomers with one chiral center and the four substituents are combinations of the groups H, F, Cl, Br, Me, Et, CN, CF_3, OH, BH_2, PH_2, AsH_2, SeH, SH, SiH_3, and Ar. Significant change in electron distribution, frontier orbitals and chemical reactivity were observed for the enantiomers in an external electric filed. Total energy difference of a chiral R/S pair under an external electric field was found to be correlated with various physiochemical properties. It was observed that the change in molecular structure and electron distribution of enantiomeric molecules in the presence of an applied electric are transformed into other changes such as frontier orbitals and regioselectivity properties.

Table 8 Riemann surfaces of R and S isomers

	R-isomer	S-isomer
Real part(sqrt)		
Imaginary part(sqrt)		
Complex map (sqrt)		
Riemann surface (sqrt)		
Real part (1/sqrt)		
Imaginary part (1/sqrt)		
Real part control plot (1/sqrt)		
Imaginary part control plot (1/sqrt)		

(continued)

Table 8 (continued)

	R-isomer	S-isomer
Complex map (1/sqrt)		
Riemann surface (1/sqrt)		

Another method of chiral distinction is by microwave three wave mixing M3WM. This method was introduced by Patterson et al. a decade ago [118, 119]. M3WM is a high-resolution spectroscopic method of chiral distinction in gas phase which is a non-linear, resonate and coherent approach. The fundamental of the method is based on the fact that not only enantiomers are mirror images of each other but also their dipole moments. The non-zero electric dipole moment components namely, μ_a, μ_b, and μ_c in a principal inertia axis system has a positive triple product i.e., $[\mu_a.(\mu_b \times \mu_c] > 0$. When two components are interchanged the triple product becomes negative i.e., $[\mu_a.(\mu_b \times \mu_c)] < 0$. Measurement of the phase difference is used as a measure of chirality and thus, the distinction of enantiomers. In the actual experiment, the chiral molecules are polarized by applying two orthogonal microwave fields. This results in two transitions in the cyclic three-level system and these are denoted as the "drive" and the "twist." The three electric dipoles μ_a, μ_b, and μ_c allow the a- b- and c- type transitions. As only two transitions are excited by applying the microwave fields the third transition is induced as a molecular response. This coherently induced transition, "listen" will be in the third mutually orthogonal polarization direction. The listen signal measured as a function of time is an indirect measurement of the phase difference of the triple product and is directly proportional to the triple product of the electric dipoles. Thus, phase difference provides a highly sensitive measurement of chirality. The details on the fundamental principles, development of the experimental technique and application are presented in seminal reviews on the topic [120–122]. Sun and Schenell [123] used six transitions instead of three transition and have reported M6WM method.

Another very recent technology in chiral distinction is microwave chiral tagging [123, 124]. In chiral tagging a small chiral molecule used as a tag to form a complex with each of the enantiomers two enantiomers of a chiral chemical and thus the enantiomers under investigation are converted into diastereomers. Once both the enantiomers are derivatized into weakly bound diastereomeric complexes with the tag the mirror image relation vanishes and the diastereomers are characterized by rotational spectroscopy. Unlike X-ray crystallography wherein assignment of absolute configuration is possible only for crystallizable solids the new method can be used for any chiral molecule.

7　Conclusion

The estimated global chiral chemical market size was USD 48 billion in the year 2021 and is expected to reach about 147 billion by 2030 at a CGAR (compound annual growth rate) of 11.25% during this period (2021–2030). The market includes pharmaceuticals, agrochemicals, flavors and fragrances [125]. The key factors for the growth are the policy decisions by FDA and EPAs and the consequent demand for the enantiopure drugs and agrochemicals, advancements in chiral separation techniques [126–129] and biocatalytic synthesis of chiral chemicals [24, 130, 131]. Under this situation many of the lead compounds in pharmaceutical and agrochemical industry are chiral. Many racemate drugs are being replaced by enantiopure single isomer drugs either by separation methods or by *de nova* synthetic methods (synthesis by design). Hence, there is a pressing need for developing predictive models for pharmacological and toxicological properties of chiral molecules. Creating all possible stereoisomers for each chiral compound in a structure library leads to an exponential increase in the number of structures to be evaluated. Though synthesis of new small molecule drugs is not attracting much attention and there is a decrease in the number of new molecules coming out of the pipeline all the chiral molecules in the use as pharmaceuticals or agrochemicals as such or as metabolites will reach the environment. Hence, the chiral descriptors will be very useful in predicting their environmental fate and soil decomposition. It is important to note that microbial decomposition of enantiomers is not the same [132, 133]. The forgoing discussion has reviewed several chirality descriptors and many of them find application in QSAR modelling. However, only a few of them can be calculated easily that too in a batch mode for a large set of chiral molecules that is very essential for screening.

　　RCI approach of the computation of chirality descriptors is not only simple based on the graph theoretic approach, but also versatile because it can use a variety of methods based on simple connectivity, valence and bond-order connectivity, information theory, electro-topological states, molecular shape index etc., to assign weights to generate a family of chirality indices for the same set of molecules. Because we cannot predict a priori what will be the qualitative or quantitative effects of the different chiral forms on the stereospecific recognitions of ligands by the numerous biological receptors a mutually different and uncorrelated as well as a broad set of chirality indices will have higher chance of finding biological correlations with the dependent variable, property, or activity. It may be noted that Kier and Hall took a similar approach of expanding the weighting of molecular graphs using different known physiochemically important molecular properties in their extension of the connectivity approach to the valence connectivity and electro topological state approaches [97, 134, 135]. It is hoped that the multidimensional method of the RCI approach will find application in diverse types of QSARs for the emerging field of chiral chemistry, pharmacology, and toxicology.

Acknowledgements The authors are grateful to Gregory D. Grunwald, Natural Resources Research Institute, Duluth, MN 55811, United States for technical support.

Dr. S. Priyarega, Department of Chemistry, Saranathan College of Engineering, Tiruchirappalli 620012, Tamil Nadu, India for her help in compiling the evolution of chirality.

References

1. Kubbinga H (2012) Crystallography from Hau¨y to Laue: controversies on the molecular and atomistic nature of solids. Z Kristallogr 227: 1–26
2. Leclercq F (2013) Arago, biot, and fresnel elucidate circular polarization. Revue d'histoire des sciences 66:395–416. https://doi.org/10.3917/rhs.662.0395
3. Aboul-Enein HY, Ali I (2004) Chiral pollutants: distribution, toxicity and analysis by chromatography and capillary electrophoresis. Wiley
4. Ramsay OB (1981) Nobel prize topic in chemistry—a series of historical monographs of fundamental chemistry—stereochemistry. Heyden & Sons Ltd.
5. Flack HD (2009) Louis Pasteur's discovery of molecular chirality and spontaneous resolution in 1848, together with a complete review of his crystallographic and chemical work. Acta Cryst A 65:371–389
6. Flake HD (2012) Perspective and concepts: chirality in nineteenth century science. Elsevier Ltd.
7. Vantomme G, Crassous J (2021) Pasteur and chirality: a story of how serendipity favors the prepared minds. Chirality 33(10):559–744
8. Drayer DE (2001) The early history of stereochemistry: from the discovery of molecular asymmetry and the first resolution of a racemate by Pasteur to the asymmetrical chiral carbon of van'thoff and Le Bel. Clin Research Reg Affairs 18(3):181–203
9. Fisher NW (1975) Wislicenus and lactic acid: the chemical background to van't Hoff's hypothesis, van't Hoff-Le Bel Centennial. Ramsay O, ACS Symposium Series, American Chemical Society, Washington, DC
10. Grossman RB (1989) Van't Hoff Le Bel, and the development of stereochemistry: a reassessment. J Chem Educ 66(1):30–33
11. Mauskopf S (2006) A history of chirality. In: Kenneth WB, Marianna AB (eds) Chiral Analysis, Elsevier, pp 3–24. https://doi.org/10.1016/B978-044451669-5/50001-6
12. Snelders HAM, Le JA (1975) Bel's stereochemical ideas compared with those of J H van't Hoff (1874), van't Hoff-Le Bel Centennial, Ramsay, O. ACS Symposium Series, American Chemical Society, Washingt on, DC
13. Ponce YM, Diaz HG, Zaldivar VR, Torrens F, Castro EA (2004) 3D-Chiral quadratic indices of the 'molecular pseudograph's atom adjacency matrix and their application to central chirality codification: classification of ace inhibitors and prediction of σ-receptor antagonist activities. Bioorg Med Chem 12:5331–5342
14. Agashe GS (1913) Stereoisomerism and optical activity, a critical study, with a new suggestion, science progress in the twentieth century (1906–1916), vol 8, no 30. Sage Publications, Ltd., pp 227–249 http://www.jstororg/stable/43432140
15. Kurihara N, Miyamot J (ed) (1998) Chirality in agrochemicals. Wiley Series in agrochemicals and plant protection. Wiley, New York
16. Silverstein RM (1998) Chirality in insect communication. J Chem Ecol 14:1981–2004

17. Leonov A, Bielory L (2007) Chirality in ocular agents. Curr Opin Allergy Clin Immunol 7(5): 418–23. https://doi.org/10.1097/ACI0b013e3282ef705b

18. Bielory L, Leonov A (2008) Stereoconfiguration of antiallergic and immunologic drugs. Ann Allergy Asthma Immunol 100(1):1–8, quiz 8–11, 36. https://doi.org/10.1016/S1081-120 6(10)60396

19. Baker GB, Prior TI (2002) Stereochemistry and drug efficacy and development: relevance of chirality to antidepressant and antipsychotic drugs. Ann Med 34:537–543

20. Lane RM, Bake GB (1999) Chirality and drugs used in psychiatry: nice to know or need to know? Cell Mol Neurobiol 19(3): 335–72. https://doi.org/10.1023/a:100997731966

21. Budău M, Hancu G, Rusu A, Cârcu-Dobrin M, Muntean DL (2017) Chirality of modern antidepressants: an overview. Adv Pharm Bull 7(4): 495–500 Dec 31 Erratum in: Adv Pharm Bull 2018 8(2):353. https://doi.org/10.15171/apb2017061

22. Howland RH (2009) Clinical implications of chirality and stereochemistry in psychopharmacology. J Psychosoc Nurs Ment Health Serv 47(8): 17–21. https://doi.org/10.3928/02793695-20090722-01

23. Nageswara Rao R, Guru Prasad K (2015) Stereo-specific LC and LC-MS bioassays of antidepressants and psychotics. Biomed Chromatogr 29(1): 21–40. https://doi.org/10.1001/bmc3356

24. Hutt AJ, O'Grady J (1996) Drug chirality: A consideration of the significance of the stereochemistry of antimicrobial agents. J Antimicrob Chemother 37:7–32

25. Kean WF, Lock CJ, Howard-Lock HE (1991) Chirality in antirheumatic drugs. Lancet 338:1565–1568

26. Mehvar R, Brocks DR, Vakily M (2002) Impact of stereoselectivity on the pharmacokinetics and pharmacodynamics of antiarrhythmic drugs. Clin Pharmacokinet 41:533–558

27. Ranade VV, Somberg JC (2005) Chiral cardiovascular drugs: an overview. Am J Ther 12:439–459

28. Vakily M, Mehvar R, Brocks D (2002) Stereoselective pharmacokinetics and pharmacodynamics of anti-asthma agents. Ann Pharmacother 36:693–701

29. Wainer IW, Granvil CP (1993) Stereoselective separations of chiral anticancer drugs and their application to pharmacodynamic and pharmacokinetic studies. Ther Drug Monit 15:570–575

30. Williams KM (1990) Enantiomers in arthritic disorders. Pharmacol Ther 46:273–295

31. Nau C, Strichartz GR (2002) Drug chirality in anesthesia. Anesthesiology 97:497–502

32. Čižmáriková R, Čižmárik J, Valentová J, Habala L, Markuliak M (2020) Chiral aspects of local anaesthetics. Molecules 12,25(12):2738. https://doi.org/10.3390/molecules25122738

33. Buda AB, Heyde TAD, Mislow K (9912) On quantifying chirality. Angew Chem Int Ed 31: 989–1007

34. Weinberg N, Mislow K (1995) A unification of chirality measures. J Math Chem 17:35–53

35. Petitjean M (2003) Chirality and symmetry measures: a transdisciplinary review. Entropy 5:271–312

36. Crippen GM (2008) Chirality descriptors in QSAR. Curr Comp Aid Drug Des 4:259–265

37. Natarajan R, Basak SC (2009) Numerical characterization of molecular chirality of organic compounds. Curr Comp Aid Drug Des 5:1–12

38. Natarajan R, Basak SC (2011) Numerical descriptors for the characterization of chiral compounds and their applications in modeling biological and toxicological activities. Curr Trends Med Chem 6(4):290–296

39. Guye P-A (1890) Influence de la constitution chimique des dérivés du carbone sur le seng et les variations de leur pouvoir rotatoire. Compt Rend Hebdom Acad Sci 110:714–716

40. Brown C (1890) On the relation of optical activity to the character of the radicals united to the asymmetric carbon atom. Proc Roy Soc Edinburgh 17:181–185

41. Boys SF (1934a) Optical rotatory power II-the calculation of rotatory power of a molecule containing four refractive radicals at the corners of an irregular tetrahedron. Proc Roy Soc London Series London A 144:675–692
42. Boys SF (1934b) Optical rotatory power I-a theoretical calculation for a molecule containing only isotropic refractive centres. Pro Roy Soc London, Series A 144:654–675
43. Ruch E (1968) Algebraic aspects of chirality phenomenon in chemistry. Acct Chem Res 5:49–56
44. Ruch E, Schönhofer A (1970) Theorie der chiralitätsfunktionen. Theor Chim Acta 19:287–294
45. Ruch E, Schönhofer A, Ugi I (1967) Die Vandermondesche determinante als näherungsansatz für eine chiralitätsbeobachtung, ihre verwendung in der stereochemie und zur berechnung der optischen aktivität. Thoer Chim Acta 7:420–432
46. Rassat A (1984) Un Critére de Classement des Systèmes Chiraux de Points à Partir de la Distance au Sens de Haussdorf. Compt Rend Hebdom Acad Sci Paris (série II) 299:53–55
47. Buda AB, Mislow KA (1992) Hausdorff chirality measure. J Am Chem Soc 114:6006–6012
48. Mezey PG (1998) The proof of the metric properties of a fuzzy chirality measure of molecular electron density clouds. J Mol Struct (THEOCHEM) 455:183–190
49. Mezey PG (1991) The degree of similarity of three-dimensional bodies: application to molecular shape analysis. J Math Chem 7:39–49
50. Mezey PG (1992) Similarity analysis in two and three-dimensions using lattice animals and polycubes. J Math Chem 11:27–45
51. Mezey PG, Ponec R, Amat L, Carbo-Dorca R (1999) Quantum similarity approach to the characterization of molecular chirality. Enantiomer 4:371–378
52. Grimme S (1998) Continuous symmetry measures for electronic wavefunctions. Chem Phys Lett 297:15–22
53. Luzanov AV, Babich EN (1992) Electronic and topological chirality indexes for dissymmetric molecular systems. Struct Chem 3:175–181
54. Luzanov AV, Babich EN (1993) Quantum-chemical quantification of molecular complexity and chirality. J Mol Srtuct (THEOCHEM) 333:279–290
55. Luzanov AV, Nerukh D (2007) Simple one-electron invariants of molecular chirality. J Math Chem 41:417–435
56. Capozziello S, Lattanzi A (2003) Geometric approach to center molecular chirality: a chirality selection rule. Chirality 15:227–230
57. Capozziello S, Lattanzi A (2003) Algebraic structure of central molecular chirality starting from fischer projections. Chirality 15:466–471
58. Capozziello S, Lattanzi A (2004) Description of Chiral tetrahedral molecule via an aufbau approach. J Mol Struct (THEOCHEM) 673:205–209
59. Zabrodsky H, Peleg S, Avnir D (1992) Continuous symmetry measures. J Am Chem Soc 114:7843–7851
60. Zabrodsky H, Avnir D (1995) Continuous symmetry measures 4 Chirality. J Am Chem Soc 117:462–473
61. Keinan S, Avnir D (1998) Quantitative chirality in structure-activity correlations shape recognition by trypsin by the D2 dopamine receptor, and by cholineesterases. J Am Chem Soc 120:6152–6159
62. Keinan S, Avnir D (1998) Quantitative chirality in structure-activity correlations shape recognition by trypsin by the d2 dopamine receptor, and by cholineesterases. J Am Chem Soc 120:6152–6159
63. Avnir D, Hel-Or HZ, Mezey PG (1998) symmetry and chirality: Continuos measures In: Schleyer PVR, Allinger NL, Clark T, Gasteiger J, Kollman PA, Schaefer HF, Schreiner R (eds) The encyclopedia of computational chemistry, vol 4. Wiley, Chichester, pp 2890–2901

64. Continuous Symmetry and Chirality Measures (CoSyM, The Hebrew University of Jerusalem (online tool for measuring the degree of symmetry, of chirality, and of polyhedral shapes). https://csm.ouproj.org.il/

65. Seri-Levy A, West S, Richards WG (1993) Chiral drug potency: Pfeiffer's rule and computed chirality coefficients. Tetrahedron Asymmetry 4:1917–1923

66. Seri-Levy A, West S, Richards WG (1994) Molecular similarity, quantitative chirality and QSAR for chiral drugs. J Med Chem 37:1727–1732

67. Benigni R, Cotta-Ramusino M, Gallo G, Giorgi F, Giuliani A, Varì MR (2000) Deriving a quantitative chirality measure from molecular similarity indices. J Med Chem 43:3699–3703

68. Basak SC, Natarajan R, Nowak W, Miszta P, Klun JA (2007) Three-dimensional structure-activity relationships (3D-QSAR) for insect repellency of diastereoisomeric compounds: a hierarchical molecular overlay approach. SAR QSAR Environ Res 18:237–250

69. Natarajan R, Basak SC, Balaban AT, Klun JA, Schmidt WF (2005) Chirality index, molecular overlay and the biological activity of diastereoisomeric mosquito repellents. Pest Manage Sci 61:1193–1201

70. Aires-de-Sousa J, Gasteiger J, Gutman I, Vidovic DI (2004) Chirality codes and molecular structure. J Chem Inf Comput Sci 44:831–836

71. Aires-de-Sousa J, Gasteiger J (2005) Prediction of enantiomeric excess in a combinatorial library of catalytic enantioselective reactions. J Comb Chem 7:298–301

72. Zhang QY, Carrera GA, Gomes MJS, Aires-de-Sousa J (2005) Automatic assignment of absolute configuration from 1D NMR data. J Org Chem 70: 2120–2130. https://doi.org/10.1021/jo048029z

73. Aires-de-Sousa J, Gasteiger J (2002) Prediction of enantiomeric selectivity in chromatography—application of conformation-dependent and conformation-independent descriptors of molecular chirality. J Mol Graph Model 20:373–388

74. Aires-de-Sousa J, Gasteiger J (2001) New description of molecular chirality and its application to the prediction of the preferred enantiomer in stereoselective reactions. J Chem Inf Comput Sci 41:369–375

75. Caetano S, Aires-de-Sousa J, Daszykowski A, Heyden YV (2005) Prediction of enantio selectivity using chirality codes and classification and regression trees. Anal Chim Acta 544:315–326

76. Zhang Q-Y, Aires-de-Sousa J (2006) New description of molecular chirality and its application to the prediction of the preferred enantiomer in stereoselective reactions. J Chem Inf Model 46:2278–2287

77. Basak, SC, Harriss, DK Magnuson, VR 1988 POLLY (Software), Copyright of University of Minnesota, USA

78. Basak SC, Magnuson VR, Niemi GJ, Regal RR (1988) Determining structural similarity of chemicals using graph-theoretic indices. Discrete Appl Math 19:17–44

79. Basak SC, Niemi GJ, Veith GD (1990) Optimal characterization of structure for prediction of properties: In: Kennedy JW, Quintas LV (eds) Special volume: Applications of graph theory in chemistry and physics. J Math Chem 4: 185–205

80. Basak SC, Restrepo G, Villaveces JL (ed) (2015) Advances in mathematical chemistry and applications, vols 1 and 2. Elsevier & Bentham Science Publishers, (27 chapters)

81. Basak SC, Vracko M (ed) (2022) Big data analytics in cheminformatics and bioinformatics (with applications to computer-aided drug design, cancer biology, emerging pathogens and computational toxicology). Elsevier.

82. Johnson M, Basak SC Maggiora G (1988) A characterization of molecular similarity methods for property prediction. Mathl Comput Modelling 11: 630–634

83. Devillers J, Balaban AT (eds) (1999) Topological indices and related descriptors in QSAR and QSPR. Gordon and Breach, Amsterdam, Netherlands
84. Karelson M (2000) Molecular descriptors in QSAR/QSPR. Wiley-Interscience, New York
85. Todeschini R, Consonni V (2000) Medicinal chemistry. In: Mannhold R, Kubinyi H, Timmerman H (eds) Weinheim, Germany, Wiley-VCH, vol 11
86. Todeschini R, Consonni V (2009) Molecular descriptors for cheminformatics, 2nd edn., vol 1, pp 127–135 Wiley-VCH, Weinheim
87. Balaban AT (1997) From chemical graphs to 3D modeling. In: Balaban AT (ed) From chemical topology to three-dimensional geometry. Plenum Publishing Corporation, New York, pp 1–24
88. Van-de Waterbeemd, H, El-Tayar N, Testa B, Wikström H, Largent B (1987) Quantitative structure-activity relationships and eudismic analyses of the presynaptic dopaminergic activity and dopamine d2 and σ receptor affinities of 3-(3-hydroxyphenyl)piperidines and octahydrobenzo[f]quinolones. J Med Chem 30: 2175–2181
89. Cramer RD III, Patterson DE, Bunce JD (1988) Comparative molecular field analysis (COMFA) 1 effect of shape on binding of steroids to carrier proteins. J Am Chem Soc 110:5959–5967
90. Marshall GR, Cramer RDIII (1988) Three-dimensional structure-activity relationships trends. Pharmacol Sci 9:285–289
91. Bucholz E, Brown RL, Tropsha A, Booth RG, Wyrick SD (1999) Synthesis, evolution, and comparative molecular field analysis of 1-phenyl-3-amino-1,2,3,4-tetrahydronaphthalenes as ligands for histamine H1 receptors. J Med Chem 42:3041–3054
92. Holloway MK, Wai JM, Halgren TA, Fitzgerald PMD, Vacca JP, Dorsey BD, Levin RB, Thompson WJ, Chen LJ, deSolms SJ, Gaffin N, Ghosh AK, Giuliani EA, Graham SL, Guare JP, Hungate RW, Lyle TA, Sanders WM, Tucker TJ, Wiggins M, Wiscount CM, Woltersdorf OW, Young SD, Darke PL, Zugay JA (1995) A priori prediction of activity for HIV-1 protease inhibitors employing energy minimization in the active site. J Med Chem 38:305–317
93. Pérez C, Pastor M, Ortiz AR, Gago F (1998) Comparative binding energy analysis of hiv-1 protease inhibitors: incorporation of solvent effects and validation as a powerful tool in receptor-based drug design. J Med Chem 41:836–852
94. Dinan L, Hormann RE, Fujimoto T (1999) An extensive ecdysteroid CoMFA. J Comput Aid Mol Des 13:185–207
95. Schultz HP, Schultz EB, Schultz TP (1995) Topological organic chemistry 9 Graph theory and molecular topological indices of stereoisomeric organic compounds. J Chem Inf Comput Sci 35: 864–870
96. de Julian-Ortiz JV, de Alapont CG, Rios-Santamarina I, Garcia-Domenech R, Galvez J (1998) Prediction of properties of chiral compounds by molecular topology. J Mol Graph Model 16: 14–18
97. Hall LH, Kier LB (1999) Molecular structure description: the electrotopological state. Academic Press, San Diego
98. Golbraikh A, Bonchev D, Tropsha A (2001) Novel chirality descriptors derived from molecular topology. J Chem Inf Comput Sci 41:147–158
99. Zheng W, Tropsha A (2000) Novel variable selection quantitative structure-property relationship approach based on the k-nearest-neighbor principle. J Chem Inf Comput Sci 40:185–194
100. Golbraikh A, Tropsha A (2003) **QSAR** modeling using chirality descriptors derived from molecular topology. J Chem Inf Comput Sci 43:144–154
101. Lukovits I, Linert W (2001) A topological account of chirality. J Chem Inf Comput Sci 41:1517–1520
102. Pyka A (1993) A new optical topological index (i_{opt}) for predicting the separation of d and l optical isomers by TLC part III. J Planar Chromatogr 6:282–288

103. Yang C, Zhong C (2005) Chirality factors and their application to QSAR studies of chiral molecules. QSAR Comb Sci 24:1047–1055
104. Diaz HG, Sánchez IH, Uriarte E, Santana L (2003) Symmetry Considerations in markovian chemicals 'In Silico' design (MARCH-INSIDE) I: central chirality codification, classification of ACE inhibitors and predictions of sigma-receptor antagonist activities. Comp Biol and Chem 27:217–227
105. Castillo-Garit JA, Ponce YM, Torrens F, García-Domenech R, Rodríguez-Borges JE (2009) Applications of bond-based 3d-chiral quadratic indices in QSAR studies related to central chirality codification. QSAR Comb Sci 28:1465–1477
106. Natarajan R, Basak SC, Neumann TS (2007) Novel Approach for the numerical characterization of molecular chirality. J Chem Inf Model 47:771–775
107. Mills JA, Klyne W (1954) Progress in stereochemistry. Butterworths, London, pp 204–212
108. Weininger D (1988) SMILES, a chemical language and information system 1 Introduction to methodology and encoding rules. J Chem Inf Comput Sci 28:31–36
109. Weininger D, Weininger A, Weininger JL (1989) SMILES 2 Algorithm for generation of unique SMILES notation. J Chem Inf Comput Sci 29:97–101
110. Natarajan R, Anbalagan TM, Murali TM (2004) INDCAL (an inhouse software for calculation of topological indices of organic molecules)
111. Balaban, AT, Mill, D, Kodali, V, Basak, SC (2005) Complexity of chemical graphs in terms of size, branching, and cyclicity, SAR QSAR Environ Chem 17: 429–466. https://doi.org/10.1080/10629360600884421
112. Filip PA, Balaban TS, Balaban AT (1987) A new approach for devising local graph invariants: derived topological indices with low degeneracy and good correlation ability. J Math Chem 1:61–83
113. Basak SC (2013) Philosophy of mathematical chemistry: a personal perspective. HYLE—Int J Philos Chem 19(1): 3–17
114. Ursu O, Diudea MV (2005) TOPOCLUJ software program. Babes-Bolyai University, Cluj
115. Ursu O, Diudea MV, Nakayama S (2006) 3D molecular similarity: method and algorithms. J Comput Chem Japan. 5: 39–46. https://doi.org/10.2477/jccj539
116. Diudea MV, Ursu O (2003) Layer matrices and distance property descriptors. Indian J Chem 42A:1283–1294
117. Chen J, Liu S, Li M, Rong C, Liu S (2020) A density functional theory and information-theoretic approach study of chiral molecules in external electric fields. Chem Phy Let 757: 137858. https://doi.org/10.1016/jcplett2020137858
118. Patterson D, Schnell M, Doyle J (2013) Enantiomer-specific detection of chiral molecules via microwave spectroscopy. Nature 497: 475–477. https://doi.org/10.1038/nature12150
119. Patterson D, Doyle JM (2013) Sensitive chiral analysis via microwave three-wave mixing. Phys Rev Lett 111(2): 023008. https://doi.org/10.1103/PhysRevLett111023008
120. Cameron RP, Götte JB, Barnett SM (2018) Chiral rotational spectroscopy. In: Polavarapu PL (ed) Chiral analysis: advances in spectroscopy, chromatography and emerging methods. Elsevier, Cambridge, USA, pp 731–752
121. Domingos SR, Pérez C, Schnell M (2018) Sensing chirality with rotational spectroscopy. Annu Rev Phys Chem 69: 499–519. https://doi.org/10.1146/annurev-physchem-052516-050629
122. Pate BH, Evangelist L, Caminati W, Xu Y, Thomas J, Patterson D, Pérez C, Schnell M (2018) Quantitative chiral analysis by molecular rotational spectroscopy. In: Polavarapu PL (ed) Chiral analysis: advances in spectroscopy, chromatography and emerging methods 2018. Elsevier, Cambridge, USA, pp 679–730

123. Sun W, Schnell M (2023) Microwave three-way mixing spectroscopy of chiral molecules weakly bound complexes. J Phys Chem Lett 14: 7389–7394. https://doi.org/10.1021/acsjpc lett3c01900

124. Mayer K, West C, Marshall FE, Sedo G, Grubbs NGS, Evangelisti L, Pate BH (2022) Accuracy of quantum chemistry structures of chiral tag complexes and the assignment of absolute configuration. Phys Chem Chem Phys 24: 27705–27721. https://doi.org/10.1039/D2CP04060C

125. https://www.globenewswirecom/news-release/2022/05/30/2452388/0/en/Chiral-Chemicals-Market-Worth-USD-147-668-2-Million-by-2030-at-11-25-CAGR-Report-by-Market-Res earch-Future-MRFRhtml

126. Cardoso PA, César IC (2018) Chiral method development strategies for HPLC using macrocyclic glycopeptide-based stationary phases. Chromategraphia 81: 841–850. https://doi.org/10.1007/s10337-018-3526-0

127. Parker D (1991) NMR determination of enantiomeric purity. Chem Rev 91:1441–1457

128. Williams ML, Wainer IW (2002) Role of chiral chromatography in therapeutic drug monitoring and in clinical and forensic toxicology. Ther Drug Monit 24:290–296

129. Yamaguchi S (1983) Asymmetric synthesis. In: Morrison JD (ed) Academic Press, New York, vol 1, Chapter 7, p 125

130. Zhang ZJ, Pan J, Ma BD, Xu JH (2014) Efficient biocatalytic synthesis of chiral chemicals. In: Ye Q, Bao J, Zhong JJ (eds) Bioreactor engineering research and industrial applications I. Advances in biochemical engineering/biotechnology, vol 155. Springer, Berlin, Heidelberg. https://doi.org/10.1007/10_2014_291

131. Zheng GW, Xu JH (2011) New opportunities for biocatalysis: driving the synthesis of chiral chemicals. Curr Opin Biotechnol 22(6): 784–92. https://doi.org/10.1016/jcopbio201107002

132. Elder FCT, Feil EJ, Pascoe B, Sheppard SK, Snape J, Gaze WH, Kasprzyk-Hordern B (2021) Stereoselective bacterial metabolism of antibiotics in environmental bacteria—a novel biochemical workflow. Front Microbiol 12: 562157. https://doi.org/10.3389/fmicb2021562157

133. Moreira IS, Ribeir, AR, Afonso CM, Tiritan ME, Castro PM (2014) enantioselective biodegradation of fluoxetine by the bacterial strain *Labrys portucalensis* F11. Chemosphere 111: 103–111. https://doi.org/10.1016/jchemosphere201403022

134. Kier L B, Hall LH (1976) Molecular connectivity in chemistry and drug research. Academic Press, New York

135. Kier LB, Hall LH (1986) Molecular connectivity in structure-activity analysis. J Wiley & Sons, New York

QSAR Modeling Using Molecular Fragment Descriptors

Suman K. Chakravarti

Abstract

Computationally generated molecular fragments offer an effective way to represent essential molecular properties, especially when detailed mechanistic information is limited. These fragments provide a means to categorize molecules into different classes by encoding functional groups responsible for specific chemical, physical, toxic, or biological characteristics. In drug discovery, utilizing libraries of small fragments is more efficient than whole molecules due to their reduced number compared to drug-like compounds. Molecular fragments serve both educational and practical purposes, aiding in understanding structure–activity relationships and facilitating the development of interpretable QSAR models for virtual screening and regulatory decision making. Despite their simplicity, the computational representation and handling of molecular fragments involves sophisticated techniques. This chapter describes common concepts, various fragment representations, and illustrates their application in building QSAR/QSPR models through examples.

Keywords

QSAR • QSPR • Molecular fragments • Molecular graphs • Distributed representation • Substructure search • Topological indices • Molecular descriptors • Machine learning

S. K. Chakravarti (✉)
MultiCASE Inc, 5885 Landerbrook Dr. #210, Mayfield Heights, OH 44124, USA
e-mail: chakravarti@multicase.com

1   Molecular Fragments: Building Blocks of Chemistry

Molecular fragments are building blocks of chemical structures [1]. They consist of smaller groups of atoms that can be combined to create more complex structures. Although the full structure of a molecule makes it unique, molecular fragments generalize the molecule and help group it into different classes. In drug discovery, exploring chemical space is more efficient using libraries of small fragments rather than larger molecules, as the number of potential small fragments is much less compared to the number of drug-like molecules [2, 3]. These fragments can represent functional groups that are responsible for specific chemical, physical, toxic, or biological properties of molecules. For example, a carboxylic acid group gives acidic property, a long alkyl chain is considered lipophilic, a nitrosamine group makes a molecule a potent carcinogen, and a benzodiazepine ring system gives sedative properties. In other words, fragments succinctly describe the core properties of a molecule, similar to how keywords represent the main ideas in a text. They are relevant in 2D as well as in 3D representation of a molecule, as the specific arrangement of various fragments in space may represent a pharmacophore [4].

2   Methods for Generating Molecular Fragments

Generating molecular fragments can be accomplished through a variety of means, such as computational methods utilizing graph theoretical algorithms or physical techniques, such as breaking a molecule using ionizing radiation, as done in mass spectrometry [5]. Mass spectrometry produces ionized fragments for identification of a molecule, while computationally generated fragments can be used for evaluating the molecule's physicochemical or pharmacological properties.

Computational methods for generating molecular fragments often involve the use of graph traversal algorithms. These algorithms represent the molecular structure as a graph, i.e., a collection of vertices and edges, and navigate through it in a specific manner, starting at a selected vertex and visiting neighboring vertices before stopping. By altering the starting point, multiple fragments can be generated. The result is a collection of fragments, each a smaller subgraph of the original molecule and composed of a connected set of vertices.

Using graph traversal methodology, several types of fragments can be generated that are useful in cheminformatics and QSAR (Fig. 1), such as:

1. Linear path fragments: These fragments are composed of atoms that form a linear graph or a path graph.
2. Branched fragments: Linear fragments with one or more branches on non-terminal atoms [6].

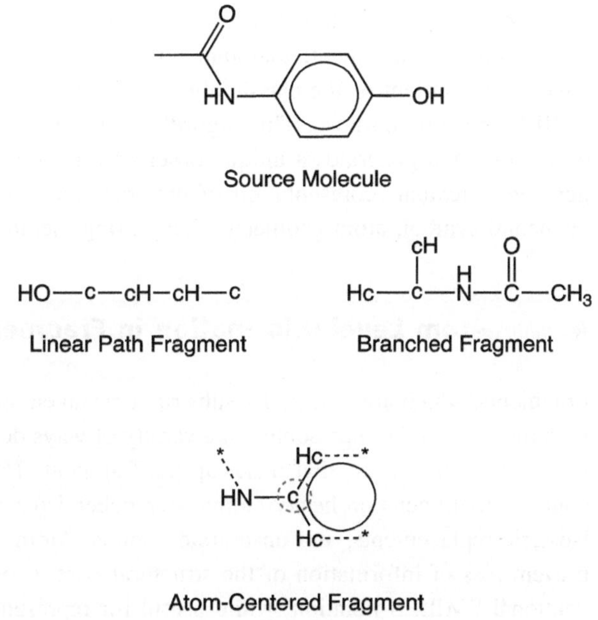

Fig. 1 Three different types of fragments generated from a molecule

Source Molecule

HO—c—cH—cH—c

Linear Path Fragment

Branched Fragment

Atom-Centered Fragment

3. Atom-centered fragments: Composed of a central atom and its successive neighboring atoms up to a certain depth [7].
4. Arbitrary shape fragments: Generated by selecting a central atom as a nucleus and adding a specified number of randomly selected neighboring atoms.

Another less commonly used but useful method for fragment generation is matching the molecule against a pre-populated dictionary of molecular fragments to identify which fragments are present in the molecule. This dictionary typically contains a large number of fragments of interest from a biological activity, toxicity or chemistry perspective, and are usually chosen by experts.

3 Unique Representation of Computationally Generated Molecular Fragments

Obtaining fragments computationally through algorithms is not enough to make them useful. Fragments in their raw form are simply collections of atoms and their bonding information. In some cheminformatics tasks, atoms are sometimes represented as integer numbers, as indices from the parent molecule. This means that the same fragment produced by two different molecules, whose atoms are numbered differently, will appear different. This poses a problem for storing the fragments in a database or using it as a descriptor in a QSAR project. To overcome this, a unique representation for every fragment must be created. However, this is a challenging task that requires encoding of structural environment, maintain human readability, and unique representation. We have used a variant of the classic algorithm published by Weininger et al. [8] for unique SMILES for this purpose. This algorithm takes the fragment graph and information for each atom, and generates a unique ordering for the atoms. This ordering can be used to generate a textual representation of the fragment. Atom level information may include elemental symbol, atom geometry, charge, ring membership, etc.

4   Atom Level Information in Fragments

Fragments, which are essentially substructures taken out of the context of a larger molecular structure, can be represented in a variety of ways depending on the type of information included for the atoms that make up the fragment. This may include details such as the count of hydrogens on heavy atoms, stereochemistry, atom geometry, ring memberships, isosteric replacements, and unsaturated bonds. Atom-level details are crucial in order to prevent loss of information of the structural context of the fragment, which is why conventional SMILES notation, while useful for representing whole molecules, may not be sufficient for representing fragments. Following are some examples of fragments with varying atom level information:

Explicit hydrogens on all atoms: [C2]([O2])([C3H2][C3H2])[O3H]
Hydrogen suppressed atoms: [C2](=[O2])(-[C3])-[O3]
Stereo   information   on   atoms: [C3_r]([C3_s])([C3][O3][P3_1]([O2])([O3H])[O3H])[O3]
Isosteric representation for atoms: [C2]([O,S,NH])([C3][C3][C3][C3][C3])[O,S,NH]
Fragment in SMILES style: OC(Nc(cc)cc)C(N)C

C2 is a carbon atom with sp2 hybridization, O2 and O3 are oxygen atoms with sp2 and sp3 hybridization respectively, C3H2 is a carbon atom with sp3 hybridization and two attached hydrogen atoms, C3_r and C3_s are carbon atoms with sp3 hybridization and R or S stereo configuration respectively, and P3_1 is a phosphorus atom with tetrahedral geometry and one attached double bond.

5   Distributed Representation of Fragments

Traditionally, molecular fragments are treated as discrete symbols when used in various applications such as fingerprints or as descriptors in QSAR. These representations are known as "one-hot" vectors, where each fragment is represented by a single active bit

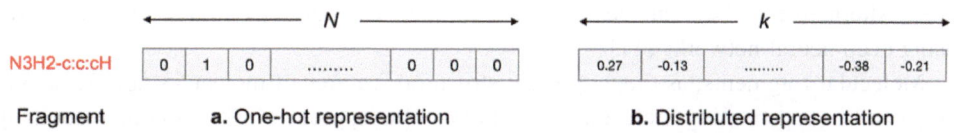

Fragment   **a. One-hot representation**   **b. Distributed representation**

Fig. 2 Two distinct ways for representing a molecular fragment: **a** the sparse and one-hot representation in which a single element of the vector is active with N total fragments in the fragment dictionary, and **b** the distributed and dense representation, with a fragment embedding size of k [9]

in a large vector, with the position of the bit indicating the fragment. However, this representation does not convey any information about the relationship between fragments. For example, using a "one-hot" vector, it is impossible to compute the similarity between the fragments $-CH_2-Br$ and $-CH_2-Cl$. On the other hand, distributed representations of fragments represent them in a high-dimensional vector space, where each fragment is represented by a multi-dimensional vector filled with continuous values, instead of a single "on" bit in a vector filled with zeroes. These distributed representations are generated through a process called "word embedding" using neural networks [9]. A schematic representation of the two types of representations is shown in Fig. 2. It is noteworthy that the size of the fragment dictionary, N, can be significantly larger than the embedding size, k, for example, hundreds of thousands compared to a few hundred.

6 Molecular Fragments in Cheminformatics, QSAR, and Drug Discovery

Molecular fragments are a familiar concept to chemists as they directly represent chemical structural features. They are widely used in a variety of applications. Fragment-based drug discovery is a field dedicated to finding new leads by starting with small, low molecular weight organic fragments that have weak binding affinity for biological targets and then growing and combining them to produce leads with higher affinity. In cheminformatics, fragments are commonly used for constructing molecular fingerprints, which are used for search and activity prediction purposes. In QSAR, fragments are primarily used as molecular descriptors to build models and predicting molecular properties.

7 Fragments and Molecular Fingerprints

Molecular fingerprints are numerical encodings of molecules and are utilized for searching similar molecules or comparing structures. They take the form of vectors, comprised of binary indicators indicating the presence or absence of various molecular characteristics, or a sequence of numerical values representing various physical and chemical properties

[10]. Alternatively, they can be generated through embedding of molecular structures using deep neural networks [11].

Molecular fragments, as they represent structural features of molecules, are frequently employed in constructing molecular fingerprints. One approach involves computationally generating molecular fragments, converting them to unique representations, and then generating a fingerprint by assigning them to specific bit positions (Fig. 3). However, the number of unique fragments produced from a set of molecules is not always known in advance and may be quite large, making it difficult to determine the appropriate length of the fingerprint. As a result, it is common to create a fixed-size fingerprint and then use a hashing algorithm to place the fragments into it. This can result in multiple distinct fragments being placed in the same bit position, a phenomenon known as bit collision.

An alternative method for constructing fingerprints involves comparing a molecule to a pre-determined set of fragments, referred to as "keys," and activating the corresponding bits in the fingerprint depending on which fragments are present in the molecule. The size of the fingerprint is predetermined and is equivalent to the size of the fragment dictionary, eliminating the possibility of bit collision. Several widely used fingerprints fall into this category, such as MACCS (166 bits) [12] and PubChem fingerprint (881 bits).

In some instances, specialized fingerprints are developed to represent the structural environments of toxicity alerts [13], as structural environments can influence their toxicity. Fragments are generated from the vicinity of the alert and the rest of the structure is disregarded, and a fingerprint is then created (Fig. 4). These fingerprints are highly useful for comparing toxicity alerts, e.g., mutagenicity alerts, in different molecular structural environments.

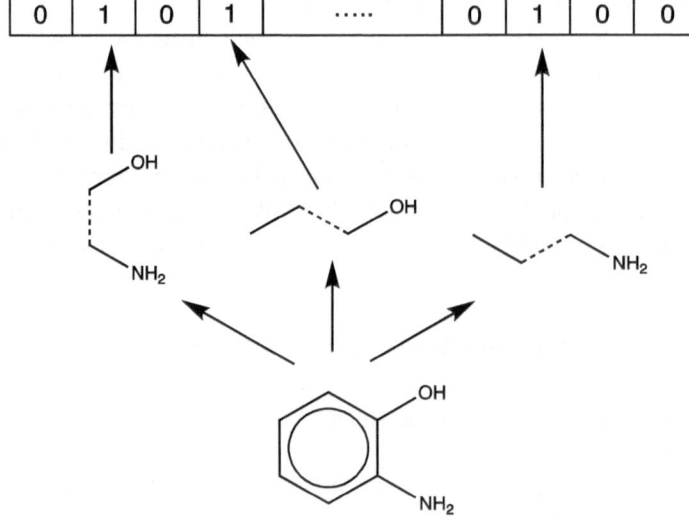

Fig. 3 The process of generating binary bit-based molecular fingerprints using molecular fragments, dotted lines in the fragments represent aromatic bonds

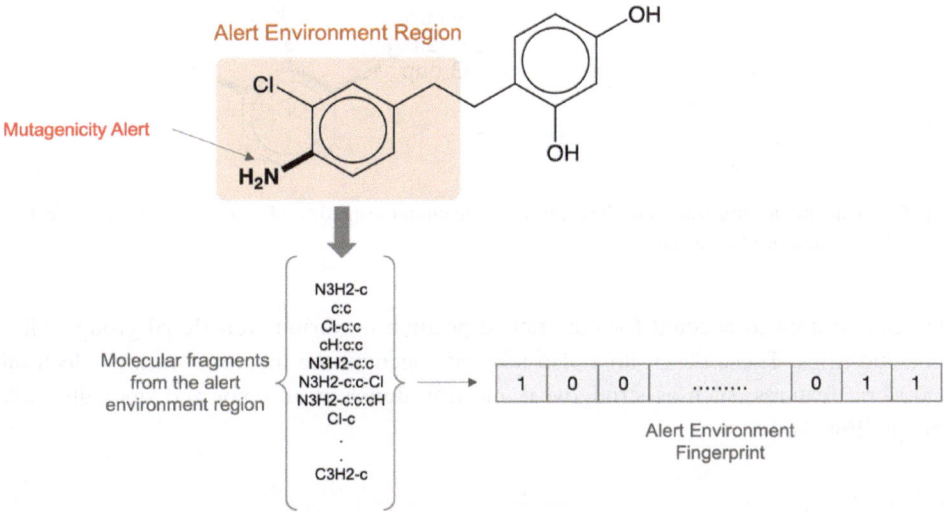

Fig. 4 A fingerprint being generated from the fragments in the region around a mutagenicity structural alert

8   Topological and Fragment Descriptors

Topological descriptors, as their name implies, are 2D graph theoretical descriptors that describe the topological characteristics of a molecule, specifically the connectivity patterns of atoms, or the atomic arrangements of molecules. They take into account branching, ring systems, size, shape, presence of heteroatoms, and bonding patterns. There are many different types of topological indices [14], such as the Wiener index [15], Hosoya index [16, 17], Estrada index [18], Randic index [19], Zagreb index [20], among others. These indices typically produce a single numerical value for the entire molecule.

On the other hand, fragment descriptors represent the presence of specific functional groups or substructures of interest in a molecule.

9   Fragment Descriptors for Aromatic Rings

Evaluating aromatic rings is crucial in modeling toxicity endpoints, e.g., for mutagenicity. However, they pose a unique challenge when it comes to utilizing fragment descriptors to accurately describe them. For instance, representing an aromatic amino group that is surrounded by two large substituents on either side can be difficult with traditional fragment descriptors. A linear fragment may not capture all the important features, while an atom-centered fragment might include atoms that are not involved in the core feature of interest. To address this issue, we have developed specialized fragment descriptors

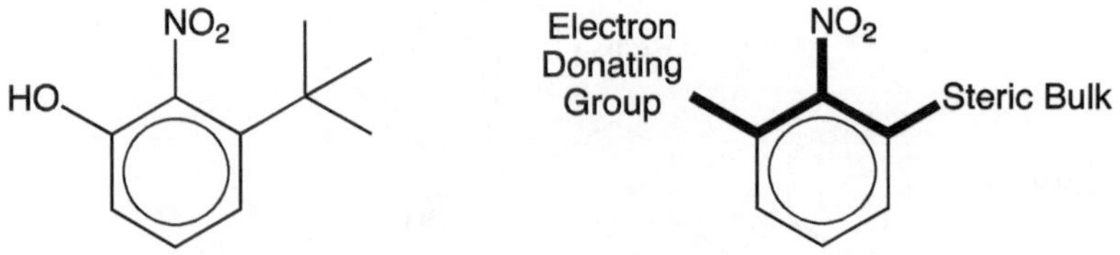

Fig. 5 An aromatic ring fragment descriptor for the characterization of electronic and steric features around an aromatic nitro group

that can be used to account for the relative position of various functional groups within aromatic rings. These descriptors also take into account the nature of various electronic and steric features, such as strong/weak electron donor or acceptor and sterically bulky groups (Fig. 5).

10   Substructure Searching Versus Fragment Matching

In many practical applications, such as QSAR and similarity searches, it is often necessary to detect the presence of a specific fragment in a query molecule. For example, when using a fragment-based QSAR model to evaluate a chemical, the first step is to determine which of the model's fragments are present in the query molecule. To do this, one can either perform a substructure search or generate all possible fragments from the query and compare them to the model's fragments to detect matches. The substructure search technique is highly flexible and can be used with fragments of any shape or size, while the comparison technique is only effective if the model fragments were generated using a specific technique, such as linear or atom-centered fragments with a specific bond depth. However, substructure searches are typically much more computationally intensive compared to fragment comparison, particularly when searching large datasets or using large substructures.

11   Fragments as Descriptors for QSAR/QSPR

Molecular fragments are frequently utilized in QSAR/QSPR models [21] for several reasons:

1. Fragments directly represent molecular structural features.
2. The models constructed using them and the evaluation outcomes are transparent and interpretable.
3. They are highly suitable for building models from large, diverse training sets.
4. These models are easy to maintain and are scalable.

5. Fragments can be used with a wide range of modeling techniques, including regression, Naive Bayes, random forests, and deep neural networks.
6. They are compatible with other descriptors, such as physico-chemical properties, E-State, molecular volume, and surface descriptors.
7. They are very useful for automatic discovery of toxicity alerts from molecular data sets.
8. Generating fragments and building models from them is fast and does not require much expertise from the modeler.

Fragment descriptors have some drawbacks too, such as:

1. Poor performance when applied to small or congeneric training sets.
2. The need of advanced variable selection techniques to construct models with desirable characteristics when working with large training sets.
3. Need of expert knowledge for the interpretation of prediction results in some cases.
4. Potential of producing statistically significant yet mechanistically meaningless toxicity alerts when using fragments to build statistical models.
5. The possibility of receiving multiple alerts for a single toxic functionality.

Creating models using fragment descriptors involves many similar steps to any other descriptor types, however, there are certain distinctions. For instance, since each training compound can generate numerous fragments as opposed to a single numerical value, a robust database system for managing fragments is necessary. Additionally, a moderate-sized training dataset can result in a large number of fragments and the number of fragments may surpass the number of training compounds, making it important to apply a good variable selection method or dimensionality reduction technique to produce a model with desirable characteristics.

The following are some steps involved in constructing a regression model utilizing fragment descriptors:

1. *Construction of training and test sets*: Compounds and their corresponding properties or activities are compiled to form a training set for model development and a test set for validation.
2. *Generation of initial fragment set*: Molecular fragments are derived from the compounds in the training set and a fragment database is created.
3. *Creation of the X-matrix*: An X-matrix of dimensions N x M is generated, with each of the N rows representing a training compound and each of the M columns representing a fragment. The non-zero matrix entries indicate the presence of a fragment in a particular training compound.
4. *Variable selection*: The X-matrix and the property values (Y values) are subjected to a variable selection process to identify a subset of relevant fragments. The variable

selection process may also generate the regression coefficients, otherwise, they must be computed by performing a regression fitting step between the selected fragments and the Y values.

5. *Model validation*: Models are validated through internal cross-validation (CV) and/ or external validation. Cross-validation is achieved by randomly withholding a percentage of the available compounds, reconstructing the model using the remaining compounds, and then evaluating the withheld compounds. This process is repeated several times, each time with a different combination of test and training sets. The model is completely rebuilt in each CV cycle, including regenerating the initial fragment set, building X and Y matrices, performing variable selection, and reweighting the descriptors. This is essentially a rigorous leave-many-out (LMO) CV process. Typically, 10%-out-10 times or 5%-out-20 times CVs are performed.

6. *Hyperparameter tuning*: Models are often associated with several hyperparameters, which can be adjusted to produce multiple versions of the model with varying performance. The hyperparameter space can be explored to identify the optimal version of the model. For example, different parameters in fragment representations may affect model performance, such as the level of detail on atoms. The modeling methodology itself can also be a hyperparameter.

7. *Building an activity/property database*: Once the optimized parameters of a model are identified and a final version of the model is constructed, a simple database containing the structures of all compounds with available experimental properties or activities is compiled and stored with the model. This database provides experimental data when available during the evaluation phase.

12   Examples of Various QSAR/QSPR Models Built Using Fragment Descriptors

The ease of use and effectiveness of molecular fragments make them suitable for constructing a wide range of models for properties, including physical and chemical, ADME, pharmacological, and ecotoxicity and adverse effects. Some specific endpoints that we have modeled are listed in Table 1. These models were constructed utilizing the FlexFilters methodology [22], an integral part of the commercial QSAR Flex software [23].

13   Applying a Fragment-Based Model for Prediction

When a model based on fragment descriptors is applied to a query compound, the first step is to check if the query is already present in the model's property database. If it is, the experimental activity/property is returned. If not, the query is then evaluated for domain applicability. If the query passes this test, it is then subjected to activity/property

Table 1 Various fragment-based QSAR/QSPR models and their performance

Module	Training set size	Test set size	Modeling method[a]	Cross validation performance[b]		Test set performance		Descriptors used[c]
				r^2/AUC	RMSE/ accuracy	r^2/AUC	RMSE/ accuracy	
Physicochemical properties								
LogP	12,645	1405	*OLSR*	0.931	0.482	0.936	0.469	*FRAGS*
Water solubility	3800	422	*RF_REGR*	0.893	0.722	0.895	0.794	*FRAGS, LOGP, MW*
Vapor pressure	1829	203	*OLSR*	0.901	1.099	0.940	0.939	*FRAGS, MW, HBA, HBD, WS, VOL_ DESCR, SURF_ DESCR, ESTATE*
Henry's law constant	531	59	*OLSR*	0.896	0.700	0.849	0.890	*FRAGS, LOGP, WS, VP*
Boiling point	4890	543	*OLSR*	0.923	23.781	0.915	26.077	*FRAGS, MW, VP*
Melting point	7787	865	*OLSR*	0.744	50.060	0.711	52.827	*FRAGS, LOGP, WS, BP, VP*
Ecotoxicity								
Bio-concentration factor	563	62	*OLSR*	0.820	0.595	0.847	0.552	*FRAGS, LOGP, WS, VP*
Fathead minnow toxicity	931	103	*OLSR*	0.662	0.817	0.686	0.941	*FRAGS, LOGP, WS, MW*

(continued)

Table 1 (continued)

Module	Training set size	Test set size	Modeling method[a]	Cross validation performance[b]		Test set performance		Descriptors used[c]
				r²/AUC	RMSE/accuracy	r²/AUC	RMSE/accuracy	
ADME								
Skin permeation	244	27	OLSR	0.856	0.377	0.882	0.310	*FRAGS, LOGP, HBA, SURF_DESCR, ESTATE*
PAMPA effective permeability	171	19	OLSR	0.649	0.721	0.605	0.624	*FRAGS, LOGP, HBD, SURF_DESCR, ESTATE*
Caco-2 permeability	1167	129	OLSR	0.471	0.679	0.490	0.716	*FRAGS*
Microsomal stability	2276	252	LOGIST_REGR	0.802	0.756	0.832	0.786	*FRAGS, LOGP, HBD, ESTATE*
Miscellaneous end points								
Skin sensitization, binary (GPMT)	1115	123	RF_CLASSF	–	0.750	–	0.846	*FRAGS*
Skin sensitization, 3-class (LLNA)	854	209	RF_CLASSF	–	0.762	–	0.713	*FRAGS, LOGP, MW, WS, BP, VP, MP*

(continued)

Table 1 (continued)

Module	Training set size	Test set size	Modeling method[a]	Cross validation performance[b]		Test set performance		Descriptors used[c]
				r^2/AUC	RMSE/ accuracy	r^2/AUC	RMSE/ accuracy	
Ototoxicity	2376	264	RF_CLASSF	–	0.765	–	0.716	FRAGS, LOGP, MW, WS, BP, VP, MP

[a] *OLSR*—Ordinary least square regression, *LOGIST_REGR*—Logistic regression, *RF_REGR*—Random Forest regressor, *RF_CLASSF*—Random Forest classifier

[b] r^2 and *RMSE* is applicable to regression methods such as *OLSR* and *RF_REGR*, *AUC* and *Accuracy* is applicable to classification methods such as *RF_CLASSF* and *LOGIST_REGR*

[c] *FRAGS*—Atom centered molecular fragments, *LOGP*—Octanol–water partition coefficient, *WS*—Water solubility, *MW*—Molecular weight, *HBA*—Count of hydrogen bond acceptors, *HBD*—Count of hydrogen bond donors, *SURF_DESCR*—Surface descriptors, *VOL_DESCR*—Volume descriptors, *ESTATE*—E-state descriptors, *VP*—Vapor pressure, *BP*—Boiling point, *MP*—Melting point

calculation, which involves matching all the model fragments with the query, and identify which fragments are present in the structure. Once these fragments are identified, their regression coefficients are used to calculate the final outcome. In the case of toxicity models that are based on alerts, such as mutagenicity, the identified fragments also serve to highlight important biologically relevant substructures in the query molecule. An example of applying the fragment-based LogP model is shown in Table 2.

Table 2 Using a fragment-based QSPR model built with simple regression method to predict

octanol–water partition coefficient of ibuprofen

Ibuprofen
Cas # 15,687-27-1
Model Name: LogP (ordinary least square regression)
Predicted LogP: 3.524
Experimental LogP: 3.97

Identified fragments in the query compound[a]	Regression coefficient
[O2]	−0.297
[C3]	0.415
[O3H]	−0.354
[C3]-[C3]	0.319
[C3]-[C3](-[C3])-[C3]	0.079
[c]	0.979
[c]-[C3]-[C3]	0.068
[c]:[c](:[c])-[C3]	0.005
[c]1:[c]:[c](:[c]:[c]:[c]1-[C3])-[C3]-[C3]	0.021
[c]-[C3](-[C2](=[O2])-[O3H])-[C3]	0.050
[c]-[C3]-[C3](-[C3])-[C3]	0.216
[c]:[c](:[c])-[C3]-[C3](-[C3])-[C3]	0.015
[c]:[c]:[c](:[c]:[c])-[C3](-[C2](=[O2])-[O3H])-[C3]	0.341

[a] [O2]—sp2 oxygen, [C3]—non-aromatic sp3 carbon, [C2]—non-aromatic sp2 carbon, [O3H]—sp3 carbon with one hydrogen, [c]—aromatic carbon, [−, = ,:]—are the symbols for single, double and aromatic bonds

14   Conclusions

In this chapter, I have emphasized a few key features of molecular fragments, particularly from the perspective of QSAR and cheminformatics. Molecular fragments are computationally easy to generate, versatile in their use for various biological or physico-chemical endpoints, and compatible with a wide range of modeling techniques. They are also well-known to practitioners from various disciplines such as chemistry, toxicology, and data science, making them highly valuable.

Not only are they useful for students in understanding the fundamental structure–activity relationships of chemical compounds, but also for building commercial QSAR/QSPR models for virtual screening and regulatory decision-making. With the growing availability of large training datasets, fragments are a reliable descriptor for QSAR/QSPR models. They provide a solid starting point which can be further enhanced by adding other descriptors, such as physicochemical properties. Additionally, the outcomes of these models provide direct insight into which parts of the query molecules are relevant to the modeled property or activity. Despite recent advancements in graph-neural networks, which use molecular structures as direct inputs, fragment descriptors remain relevant due to their simplicity and ease of use. Molecular fragments are poised to have a long future of valuable applications.

References

1. Salum LB, Andricopulo AD (2010) Fragment-based QSAR strategies in drug design. Expert Opin Drug Discov 5(5):405–412. https://doi.org/10.1517/17460441003782277
2. Congreve M, Chessari G, Tisi D, Woodhead AJ (2008) Recent developments in fragment-based drug discovery. J Med Chem 51(13):3661–3680. https://doi.org/10.1021/jm8000373
3. Sutherland JJ, Higgs RE, Watson I, Vieth M (2008) Chemical fragments as foundations for understanding target space and activity prediction. J Med Chem 51(9):2689–2700. https://doi.org/10.1021/jm701399f
4. Bajusz D, Wade WS, Satała G, Bojarski AJ, Ilaš J, Ebner J, Grebien F, Papp H, Jakab F, Douangamath A, Fearon D, von Delft F, Schuller M, Ahel I, Wakefield A, Vajda S, Gerencsér J, Pallai P, Keserű GM (2021) Exploring protein hotspots by optimized fragment pharmacophores. Nat Commun 12(1):3201. https://doi.org/10.1038/s41467-021-23443-y
5. Steckel A, Schlosser G (2019) An organic chemist's guide to electrospray mass spectrometric structure elucidation. Molecules 24(3):611. https://doi.org/10.3390/molecules24030611
6. Klopman G (1984) Artificial intelligence approach to structure-activity studies. computer automated structure evaluation of biological activity of organic molecules. J Am Chem Soc 106(24):7315–7321. https://doi.org/10.1021/ja00336a004
7. Rogers D, Hahn M (2010) Extended-connectivity fingerprints. J Chem Inf Model 50(5):742–754. https://doi.org/10.1021/ci100050t
8. Weininger D, Weininger A, Weininger JL (1989) SMILES. 2. algorithm for generation of unique SMILES notation. J Chem Inf Comput Sci 29(2):97–101. https://doi.org/10.1021/ci00062a008

9. Chakravarti SK (2018) Distributed representation of chemical fragments. ACS Omega 3(3):2825–2836. https://doi.org/10.1021/acsomega.7b02045
10. Chen X, Reynolds CH (2002) Performance of similarity measures in 2d fragment-based similarity searching: comparison of structural descriptors and similarity coefficients. J Chem Inf Comput Sci 42(6):1407–1414. https://doi.org/10.1021/ci025531g
11. Xu K, Hu W, Leskovec J, Jegelka S (2019) How powerful are graph neural networks? arXiv February 22. http://arxiv.org/abs/1810.00826. Accessed 23 Nov 2022
12. Durant JL, Leland BA, Henry DR, Nourse JG (2002) Reoptimization of MDL keys for use in drug discovery. J Chem Inf Comput Sci 42(6):1273–1280. https://doi.org/10.1021/ci010132r
13. Chakravarti SK, Saiakhov RD (2019) Computing similarity between structural environments of mutagenicity alerts. Mutagenesis 34(1):55–65. https://doi.org/10.1093/mutage/gey032
14. Basak SC, Harriss DK, Magnuson VR (1984) Comparative study of lipophilicity versus topological molecular descriptors in biological correlations. J Pharm Sci 73(4):429–437. https://doi.org/10.1002/jps.2600730403
15. Rouvray DH (2002) The rich legacy of half a century of the Wiener index. In: Topology in chemistry; Elsevier, pp 16–37. https://doi.org/10.1533/9780857099617.16
16. Hosoya H (2003) The topological index Z before and after 1971. ChemInform 34(15). https://doi.org/10.1002/chin.200315292
17. Hosoya H (1971) Topological index. A newly proposed quantity characterizing the topological nature of structural isomers of saturated hydrocarbons. Bull Chem Soc Jpn 44(9):2332–2339. https://doi.org/10.1246/bcsj.44.2332
18. Estrada E (2000) Characterization of 3D molecular structure. Chem Phys Lett 319(5–6):713–718. https://doi.org/10.1016/S0009-2614(00)00158-5
19. Randic M (1975) Characterization of molecular branching. J Am Chem Soc 97(23):6609–6615. https://doi.org/10.1021/ja00856a001
20. Das KC, Xu K, Nam J (2015) Zagreb indices of graphs. Front Math China 10(3):567–582. https://doi.org/10.1007/s11464-015-0431-9
21. Baskin I (2008) Chapter 1. Fragment descriptors in SAR/QSAR/QSPR Studies, molecular similarity analysis and in virtual screening. In: Varnek A, Tropsha A (eds) Chemoinformatics approaches to virtual screening. Royal Society of Chemistry, Cambridge, pp 1–43. https://doi.org/10.1039/9781847558879-00001
22. Chakravarti SK, Alla SRM (2023) Fast and efficient implementation of computational toxicology solutions using the FlexFilters platform. In: Hong H (ed) QSAR in safety evaluation and risk assessment. Academic Press, pp 219–234. https://doi.org/10.1016/B978-0-443-15339-6.00055-2
23. QSAR Flex. v2.6., MultiCASE Inc., (2023) https://multicase.com/qsar-flex. Accessed 24 December 2023

Quantitative Structure-Activity Analysis Using Conceptual DFT and Information Theory-based Descriptors

Arpita Poddar, Ranita Pal, Shanti Gopal Patra, and Pratim Kumar Chattaraj⊚

Abstract

Quantitative structure–activity relationship (QSAR) modeling is an important part of chemical/biological data analysis and chemoinformatics. The low cost and high speed of screening of large chemical databases, make the QSAR analysis more efficient than the experimental methods. The linear and multi-linear regression models are extensively used for predicting biological/ecotoxicological activities or properties. Various conceptual density functional theory (CDFT) and information theory-based (IT) descriptors are used to develop the QSAR models. In this regard, different experimental toxicity parameters are considered as the dependent variable, whereas some CDFT descriptors are used as the independent variables. On the other hand, IT descriptors are also used to develop the QSAR model for describing different structural parameters and properties of the chemical systems.

A. Poddar
Department of Chemistry, Indian Institute of Technology, Kharagpur 721302, India

R. Pal
Advanced Technology Development Centre, Indian Institute of Technology, Kharagpur 721302, India

S. G. Patra
Department of Chemistry, National Institute of Technology, Silchar, Assam 788010, India

P. K. Chattaraj (✉)
Department of Chemistry, Birla Institute of Technology Mesra, Ranchi, Jharkhand 835215, India
e-mail: pkc@chem.iitkgp.ac.in

S. C. Basak (ed.), *Mathematical Descriptors of Molecules and Biomolecules*, Synthesis Lectures on Mathematics & Statistics, https://doi.org/10.1007/978-3-031-67841-7_5

Keywords

Quantitative structure–activity relationship (QSAR) • Conceptual density functional theory (CDFT) • Information theory (IT) • Multi-linear regression (MLR)

1   Introduction

Quantitative structure–activity/property relationship (QSAR/QSPR) is utilized as an important tool for the pharmaceutical activity of organic molecules. Crum-Brown and Fraser in 1868 first gave the mathematical relation between 'physiological activity', Φ and structure of a chemical species, C as [1]:

$$\Phi = f(C) \tag{1}$$

Later, the linear relationship of hydrophobicity (or lipophilicity) in terms of oil–water partition coefficient, with toxicity and narcotic activity was independently shown by Richet, Meyer, and Overton [2–4]. Following the pioneering work of Hansch [5, 6], several studies [7–18] on QSAR have been reported marking its evolution through the years. To predict biological activity and toxicity, the structure–activity relationship (SAR) methods [19] are becoming increasingly popular. Various molecular properties viz., electronic, hydrophobic, etc. are used as descriptors for producing mathematical or computational models through statistical techniques, capable of predicting their activities with high accuracy. They are applied to an unknown set of molecules of similar structure to predict their activity/property/toxicity, mostly applicable in fields like pharmaceutical, ecotoxicological, drug delivery, agrochemical, cosmetic industries, etc., among many others.

The process of drug designing starts with the proper identification of the structure of the molecules (both with known and unknown activities), followed by their energy calculation, and relating the geometrical/chemical properties with the target activity of the known compounds. The model so produced is then applied to the unknown set to predict their activity. Mathematically, QSAR is represented as:

$$\text{Activity} = f(\text{physicochemical properties})$$

i.e., in the linear form,

$$\text{Activity} = a_0 + a_1 x_1 + a_2 x_2 + a_3 x_3 + \dots \tag{2}$$

where, x_n and a_n are the descriptors and the coefficients, respectively. Various statistical methods are applied for calculating the coefficients whereas the descriptors are either experimentally obtained or computationally derived.

Various mathematical algorithms have been successfully developed and applied to procure QSAR/QSPR/QSTR models [20–27]. The beginning of QSAR modeling demands the introduction of significant number of descriptors, which has become of paramount interest to evolve and enrich this field. In this regard, quantum mechanical calculations give proper descriptions of both global and local reactivity indices within the realm of conceptual density functional theory (CDFT) [28–38]. The global as well as local CDFT-based descriptors namely, orbital energies (E_{LUMO} or E_{HOMO}), electronegativity (χ) [39–41], chemical potential (μ) [41, 42], chemical softness (S) [43, 44], hardness (η) [35, 45], electrophilicity index (ω) [46] and polarizability (α) along with atomic charges (Q_k) [47], Fukui functions ($f(r)$) [48], and their condensed-to-atom variants (f_k) [49], local philicity [50], local reactivity are extensively utilized to develop regression models with respect to experimental data on hydrophobicity, IC_{50}, etc. The aforementioned molecular reactivity descriptors are employed in conjunction with some electronic structure principles, viz., maximum hardness principle (MHP) [51], minimum electrophilicity principle (MEP) [52–54], minimum polarizability principle (MPP) [55] within the domain of CDFT. According to MHP, "There seems to be a rule of nature that molecules arrange themselves so as to be as hard as possible"; whereas MPP states "The natural direction of evolution of any system is towards a state of minimum polarizability (α)". During a chemical process, molecules always tend to decrease their electrophilicity power and become less reactive as stated by MEP.

The concept of information theory and its associated descriptors provide various important physical insights. Information theory (IT) gives the quantification and communication of information for a given system employing the probability distribution function. Treating the electron density as the local distribution function, the electronic structure of atoms and molecules are determined through the IT approach. The theory to anticipate these chemical reactivity descriptors quantitatively is called density functional reactivity theory (DFRT) [30, 42, 56, 57]. In recent years, a good amount of information has been gained through IT to understand physical and chemical processes [58–72]. In this regard, the importance of electron correlation based studies have been demonstrated [73, 74]. Entropy is a focal concept in IT that quantifies the uncertainty to evaluate the value for the local distribution function. The information conservation principle is used to quantify many reactivity descriptors like regioselectivity, electrophilicity, and nucleophilicity. Various IT-based descriptors generally used to evaluate atomic and molecular properties are Fisher information [75], Shannon entropy [76], Onicescu information energy [77], and Ghosh–Berkowitz–Parr entropy [78], etc. The field of research which uses one or many of these properties to express electronic properties, total energy and components of total energy is called information functional theory (IFT) [79].

In the present book chapter, we have summarized the QSAR/QSPR/QSTR that utilizes CDFT and IT-based descriptors to describe properties of chemical systems and different structural parameters.

2   Computational Details

The quantitative definitions of global as well as local reactivity descriptors are provided by DFT-based calculation. Important descriptors particularly used in the domain of QSAR are defined as follows:

$$\mu = \left(\frac{\partial E}{\partial N}\right)_{v(\vec{r})} = -\frac{(IP + EA)}{2} = -\chi \tag{3}$$

$$\eta = \left(\frac{\partial^2 E}{\partial N^2}\right)_{v(\vec{r})} = \left(\frac{\partial \mu}{\partial N}\right)_{v(\vec{r})} \approx IP - EA \tag{4}$$

$$IP = E_{(N-1)} - E_{(N)} \tag{5}$$

$$EA = E_{(N)} - E_{(N+1)} \tag{6}$$

$$\mu = \frac{1}{2}(E_{HOMO} + E_{LUMO}) \tag{7}$$

$$\eta = -(E_{HOMO} - E_{LUMO}) \tag{8}$$

$$S = \frac{1}{\eta} = \left(\frac{\partial N}{\partial \mu}\right)_{v(\vec{r})} \tag{9}$$

$$\omega = \frac{\mu^2}{2\eta} = \frac{\chi^2}{2\eta} \tag{10}$$

$$\Delta\omega^{\pm} = \omega^+ - (-\omega^-) = \omega^+ + \omega^- \text{ or } \Delta\omega^{\pm} = \left\{\omega^+ - \frac{1}{\omega^-}\right\} \tag{11}$$

$$\omega^+ = \frac{(\mu^+)^2}{2\eta^+} = \frac{EA^2}{2(IP - EA)} \text{ OR } \omega^+ = \frac{(IP + 3EA)^2}{16(IP - EA)} \tag{12}$$

$$\omega^- = \frac{(\mu^-)^2}{2\eta^-} = \frac{IP^2}{2(IP - EA)} \text{ OR } \omega^- = \frac{(3IP + EA)^2}{16(IP - EA)} \tag{13}$$

$$\left(\frac{\delta E}{\delta v(r)}\right)_N = \rho(r) \tag{14}$$

$$f(r) = \left(\frac{\partial \rho(r)}{\partial N}\right)_{v(r)} = \left(\frac{\delta \mu}{\delta v(r)}\right)_N \tag{15}$$

$$\omega_k^{\alpha} = \omega f_k^{\alpha}; \alpha \equiv -, +, 0 \tag{16}$$

$$f_K^- = q_K(N) - q_K(N-1); \text{ for electrophilic attack} \tag{17}$$

$$f_K^+ = q_K(N+1) - q_K(N); \text{ for nucleophilic attack} \tag{18}$$

$$f_K^0 = \frac{1}{2}[q_K(N+1) - q_K(N-1)]; \text{ for radical attack} \tag{19}$$

where E, μ, and $v\,(\vec{r})$ are the energy, chemical and external potentials for an N-electron system respectively. Chemical potential (μ) and global hardness (η) are defined in terms of ionization potential (IP) and electron affinity (EA) in Eqs. 3 and 4, respectively, and in terms of orbital energies in Eqs. 7 and 8 (using Koopmans' theorem). Definitions of IP and EA are provided in terms of the total energies of neutral, anionic, and cationic (E_N, E_{N+1}, and E_{N-1}) variants of the optimized system (Eqs. 5 and 6). Global softness (S) is defined in Eq. 9 as the reciprocal of chemical hardness. Electronegativity (χ), and electrophilicity index (ω), as described by Pauling [39] and Parr et al. [46], respectively, are provided in Eqs. 3 and 10. ω possesses requisite information about the structure, stability, bonding, toxicity, reactivity, interaction, and dynamics of the molecular system. By definition, ω is associated with the lowering of energy and maximal flow of electrons between two species as given by Parr et al. While the net electrophilicity $(\Delta\omega^{\pm})$ as formulated by Chattaraj et al. [80] (Eq. 11), is used to better understand the electrophilic power of a chemical system, the electroaccepting (ω^+) and electrodonating powers (ω^-) (Eqs. 12 and 13) define the tendency of a species to accept or donate a fractional amount of charge in a certain chemical environment. Among the local CDFT descriptors, electron density $\rho(r)$ is the most important one (Eq. 14). Fukui function (FF) (Eq. 15) quantifying the change in $\rho(r)$ for an infinitesimal change in N, and its condensed-to-atom variants (Eqs. 17–19) are extensively used in understanding the local reactivities of any molecular system. q_K is the electron population at the corresponding kth site. The philicity (ω_k^α) describing the electrophilic/nucleophilic power associated with a particular site k in a molecule with α representing electrophilic $(-)$, nucleophilic $(+)$, and radical (0) attacks, is defined in Eq. 16. Besides CDFT, recently many researchers have focused more on developing an information theory-based approach in DFT and its application for predicting molecular structure and their properties. The electron density functionals like Shannon entropy (S_S), Ghosh–Berkowitz–Parr (GBP) entropy (S_{GBP}), Fisher information (I_F) and Rényi entropy (R_n) [81], Tsallis entropy (T_n) [82], Onicescu information energy of order n (E_n) are the most eminent properties in this regard. The electron density $(\rho(r))$ delocalization and spatial uniformity are measured by the Shannon entropy which is defined as:

$$S_S = -\int \rho(r) \ln \rho(r) dr = \int S_S(r) dr \tag{20}$$

On the other hand, the GBP entropy is defined by a phase-space distribution function which results from the introduction of DFT into a local version of thermodynamics.

$$S_{GBP} = \int \frac{3}{2} k\rho(r) \left[c + \ln \frac{t(r, \rho)}{t_{TF}(r, \rho)} \right] dr \tag{21}$$

where $t(r; \rho)$ is the kinetic energy density, satisfying $\int t(r; \rho) dr = T_S$. T_S is the total kinetic energy for a noninteracting system whereas $t_{TF}(r; \rho)$ is the Thomas–Fermi kinetic energy density.

Another important IT descriptor known as the Fisher information (I_F) is defined as:

$$I_F \equiv \int i_F(r) dr = \int \frac{|\nabla \rho(r)|^2}{\rho(r)} dr \tag{22}$$

where electron density gradient and local density of Fisher information are represented by $\nabla \rho(r)$ and $i_F(r)$, respectively. I_F quantifies the sharpness or concentration of the electron density distribution.

Other reactivity descriptors in ITA within the realm of DFT known as, Rényi entropy (R_n), Tsallis entropy (T_n), and Onicescu information energy (E_n) are formulated as follows.

$$R_n = \frac{1}{1-n} \ln \left[\int \rho^n(r) dr \right]; n \geq 0 \& n \neq 1 \tag{23}$$

$$T_n = \frac{1}{n-1} \left[1 - \int \rho^n(r) dr \right] \tag{24}$$

$$E_n = \frac{1}{n-1} \int \rho^n(r) dr (n \geq 2) \tag{25}$$

The other widely utilized ITA quantity is the Information gain (I_G), also known as Kullback–Leibler divergence, and is defined as:

$$I_G = \int \rho(r) \ln \frac{\rho(r)}{\rho_0(r)} dr = \sum_A \int \rho_A \ln \frac{\rho_A}{\rho_A^0} dr \tag{26}$$

where ρ_0 and $\rho_A{}^0$ are the reference electron and reference atomic densities, for ρ and atom A in a molecule, respectively.

3 Case Studies

3.1 CDFT Approach

QSAR analysis has procured great importance in toxicity prediction in the branch of pharmacological sciences. Usually, a dataset is divided into two categories, i.e., training and test sets. The regression model obtained for the training set is employed to calculate various properties of the compounds in the test set. The performance of QSAR models can be predicted by analyzing the regression coefficient (R^2) and standard deviation (SD) values. In this regard, Roy et al. [83] established the QSTR model for a set of 252 aliphatic compounds (against *Tetrahymena pyriformis*) using DFT-based descriptors, electrophilicity (ω) and local philicity (ω_m^+/ω_m^-). The regression model was formed taking the experimental toxicity ($pIGC_{50}$) as the dependent variable and the aforementioned DFT-based descriptors as independent variables. A strong correlation of $R^2 = 0.831$ is obtained between the experimental and predicted $pIGC_{50}$ values for a set of 109 aliphatic alcohols (Fig. 1a). R^2 values of 0.787 (Fig. 1b), 0.766 (Fig. 1c), 0.803 (Fig. 1d) and 0.778 (Fig. 1e) are obtained for sets of 39 aliphatic acids, 51 aliphatic esters, 13 aliphatic aldehyde, and 15 ketones, respectively. This suggests that ω and ω_m^+ can be used as reliable descriptors for explaining the toxicity of the selected compounds against *Tetrahymena pyriformis*. For 25 aliphatic amines, ω and ω_m^- give a good correlation with $R^2 = 0.791$ (Fig. 1f). The correlation coefficients calculated for 171 electron acceptor compounds and 81 electron donor compounds from the selected 252 aliphatic compounds are 0.801 and 0.870, respectively, which indicate the high predicting power of electrophilicity and philicity descriptors. One-parameter regression model using the simplest descriptor, the number of carbon atoms in the molecule (N_C), is highly proficient to describe the toxicity ($pIGC_{50}$) of *T. pyriformis* [34]. The efficacy of the model is increased when electrophilicity is used along with N_C, forming a two-parameter regression model ($R^2 = 0.9283$ and 0.8284, for the set of electron acceptors and donors, respectively; Fig. 2). Furthermore, the three-parameter regression model is achieved when local philicity and atomic charges are incorporated as the molecular descriptors along with the electrophilicity.

In another study, the toxic effects of 15 substituted benzene compounds against *Pimephales promelas* is reported [26], where ω, ω^2, log P, and (log $P)^2$ are considered as the independent variables and experimental pLC_{50} as the dependent variable. 10 molecules from 15 are considered for the training set and the rest are considered for the test set. To avoid any bias, and to make the study statistically sound [84, 85], threefold cross-validation is performed so that each and every molecule is included once in the training set. For this purpose, the entire set of data is divided into three subsets (say, A, B, and C, with 5 molecules in each of them). The training set and test set are then chosen in such a way that if the training is done with any two subsets (say, A and B), then the remaining one (i.e., C) is used for testing. The same is repeated to include all three combinations. The result shows that the one-parameter regression model using ω^2

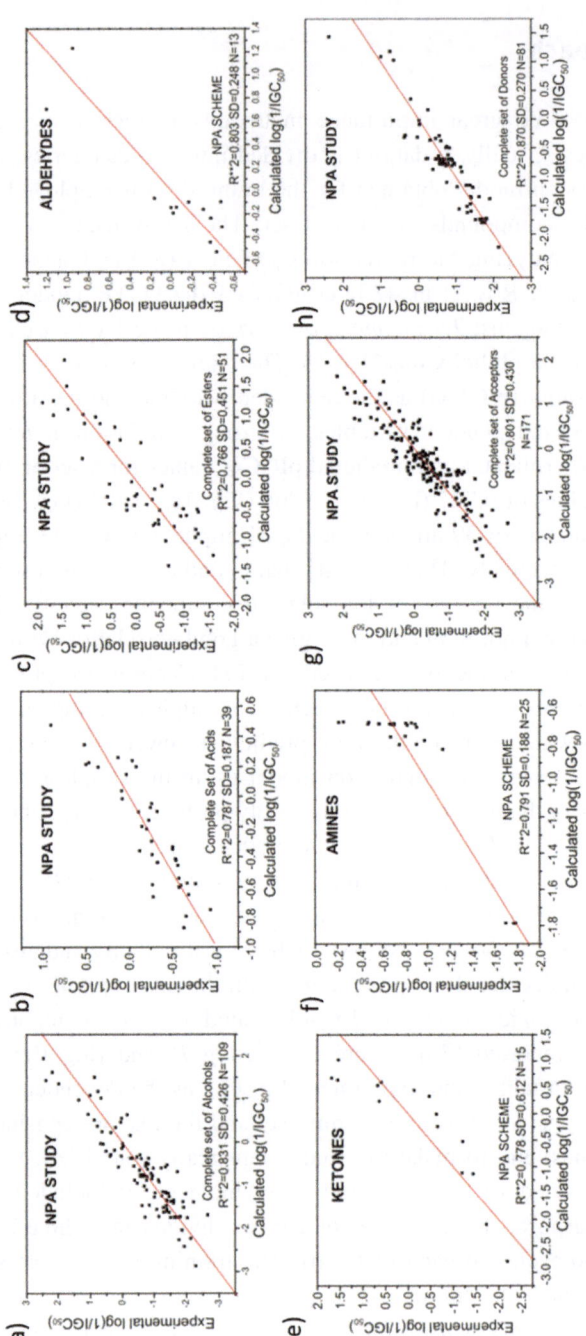

Fig. 1 Experimental versus calculated log (IGC_{50}^{-1}) values **a** all aliphatic alcohols, **b** all aliphatic acids, **c** all aliphatic esters, **d** aldehydes, **e** ketones, **f** amines, **g** electron acceptors, **h** electron donors. (Reprinted from Reference 83 with permission from Elsevier. © 2005 Elsevier Ltd. All rights reserved)

a)

b)

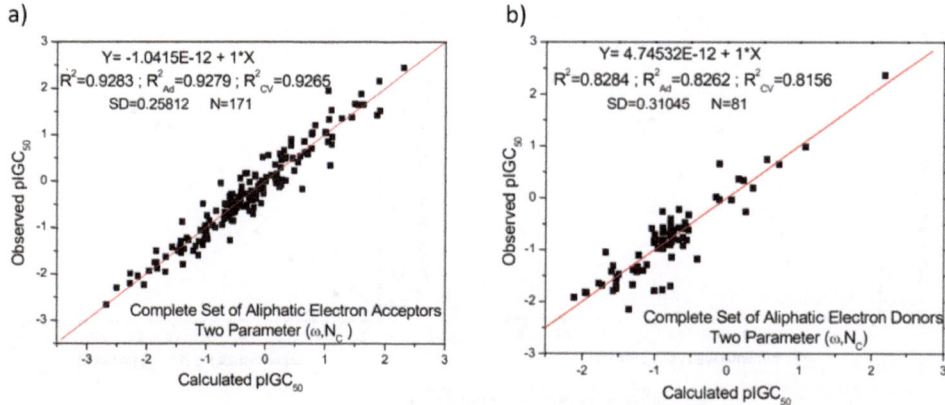

Fig. 2 Observed versus calculated pIGC$_{50}$ for the complete dataset of aliphatic electron- **a** acceptors and **b** donors. (Reprinted from Reference 34 with permission from Springer Nature. © 2008, Indian Academy of Sciences)

as descriptor has high predicting ability ($R^2 = 0.731$–0.991) than ω and it gives comparable results with log P and (log P)^2. In addition to the fact that the computation of ω is easier than the calculation of log P, the former turns out to be a better descriptor in this case. A similar study [25] reporting the toxicity of a set of 169 aliphatic compounds against *Tetrahymena pyriformis* using the electronic (ω, ω^2, ω^3) and hydrophobicity {log P, (log P)^2} descriptors reveal the combinations, {ω, log P}, {ω^2, log P} and {log P, (log P)^2}, used to build 2D QSTR models have the highest predicting powers. Electrophilicity index (ω) can also be used to predict the biological activities of male and female sex hormones like testosterone and estrogen derivatives [13]. The number of atoms in a molecule (N_A) can be used as the effective molecular descriptor to develop the QSAR model for explaining the activity of the sex hormones and the efficacy of the model increased when electrophilicity index is used along with N_A [14].

On the other hand, Padmanabhan et al. [86] reported a QSAR study on the toxicity of 19 chlorophenol (CP) compounds against *Daphnia magna*, using ω, $\omega_g{}^+$ and $\omega_g{}^-$ as the independent variables, and experimental toxicity [log (1/ EC$_{50}$)] as the dependent variable. The combination of $\omega_g{}^+$ and $\omega_g{}^-$ forming bivariate QSAR model provides the highest R^2 (0.889). The plot of observed *vs.* calculated log(1/EC$_{50}$) with $R = 0.942$ is depicted in Fig. 3a. Regression models with respect to ω, $\omega_g{}^+$ and $\omega_g{}^-$ developed against *Brachydanio rerio* and *Bacillus* are plotted in Figs. 3b and c, reveal even higher correlation coefficients ($R = 0.964$ and 0.959, respectively). QSAR study on the HAT activity of 32 pyridyl benzamide derivatives against *Trypanosoma brucei* is also reported [87] using descriptors like ω, ω^2, GATS8c, RDF40p and RDF55s.

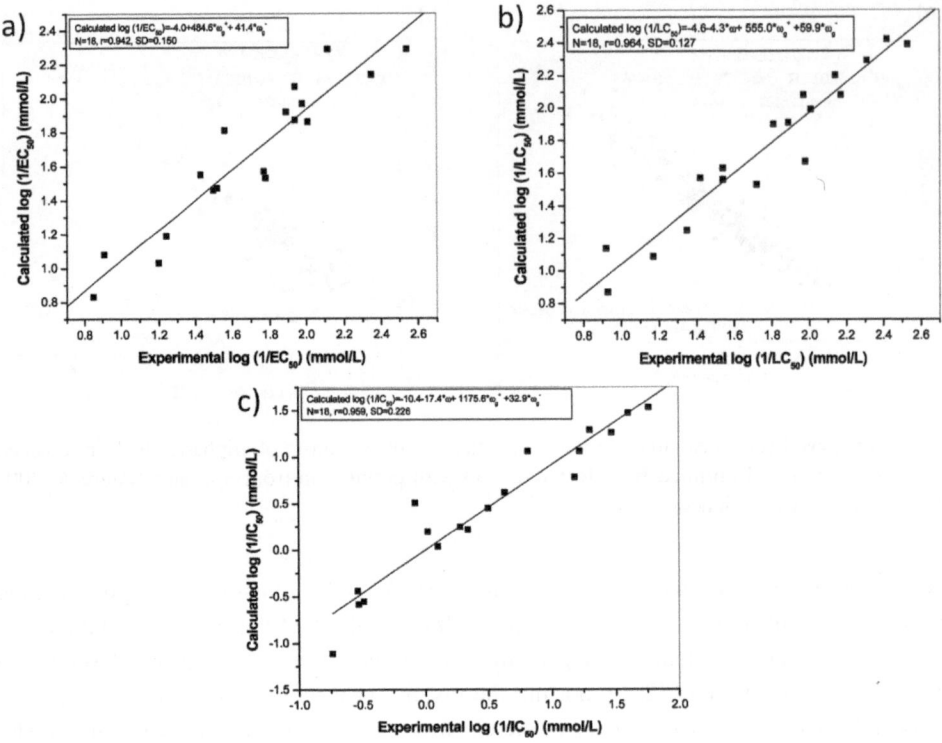

Fig. 3 Experimental versus calculated toxicity of CPs against **a** *D. magna*, **b** *B. rerio*, and **c** *Bacillus*. (Reprinted from Reference 86 with permission from American Chemical Society. © 2006 American Chemical Society)

3.2   ITA Approach

Apart from CDFT, recently researchers are greatly involved to investigate different physic-ochemical properties of a large variation of inorganic, organic and biological compounds within the realm of IT approach. The behaviour of chiral molecules in the presence of an applied external electric field employing information gain and Shannon entropy as the descriptors has been investigated from the viewpoint of ITA [68]. A strong correlation with $R^2 = 0.901$ (Fig. 4a) is observed between the total energy difference (ΔE, between R and S conformers of 74 pairs of molecules) and the total information gain difference (ΔI_G) when the external electric field is 0.005 a.u. along the x direction. I_G plays a crucial role in quantifying the electrophilicity, nucleophilicity and regioselectivity of different systems. By changing the strength of the applied electric field from 0.0 a.u. to 0.02 a.u. with step size 0.0005 a.u., the structure and distribution of electrons in the molecule CHClBrMe changes which leads to the change in information gain. A strong correlation is noticed between ΔE and other IT descriptors (Figs. 4b, c, d). The results reflect the importance of IT descriptors for describing the chemical properties and reactivities of chiral molecules in presence of an external electric field. Another physicochemical property was introduced using the IT indices within the premise of DFRT to investigate the origin of the anomeric effect of 45 systems with a general formula $R_1–X–CH_2–R_2$, where X is an electronegative heteroatom, viz., O, S, Se and R_1 and R_2 are functional groups like $–Me$, $–F$, $–Cl$, $–NH_2$, $–OH$, $–OMe$, $–CH = CH_2$, $–C≡CH$ [72]. A good correlation ($R^2 = 0.67$) exists between the energy difference (ΔE) and I_G calculated for the anti-gauche conformer of $CH_3CH_2OCH_3$ and FCH_2OF. Employing all the IT quantities i.e., S_S, I_F, E_n, S_{GBP}, I_G, R_n in the multi-linear regression analysis gives a very strong correlation coefficient of 0.99. The results demonstrate that using all IT quantities simultaneously, instead of one-at-a-time, can provide a much better fitting for this study.

Other molecular properties like molecular acidity within the realm of DFRT framework is investigated recently by Cao et al. [69]. Five different categories of acidic series are examined using IT quantities such as S_S, I_F, E_n, S_{GBP}, I_G, R_n to predict their experimental pKa values. Figure 5 shows a strong correlation of various IT quantities with the experimental pK_a values of singly substituted benzoic acid derivatives. Figure 5b has two

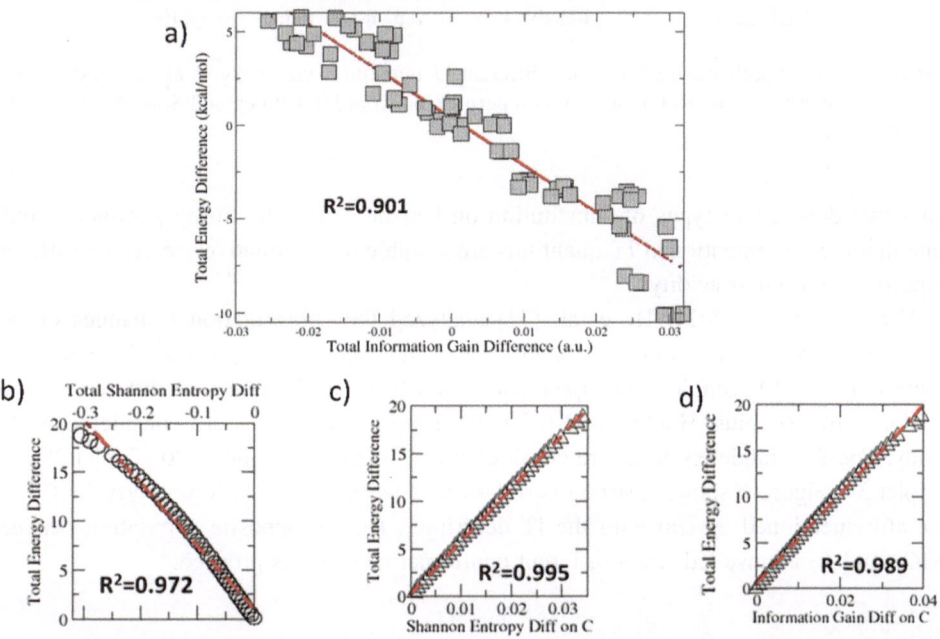

Fig. 4 Difference in total energy (ΔE, between R/S) versus **a** the difference in total information gain (ΔI_G) **b** the difference in total Shannon entropy (ΔS_S), **c** ΔS_S on the central C atom, **d** ΔI_G on the central C atom. (Reprinted from Reference 68 with permission from Elsevier. © 2020 Elsevier B.V. All rights reserved)

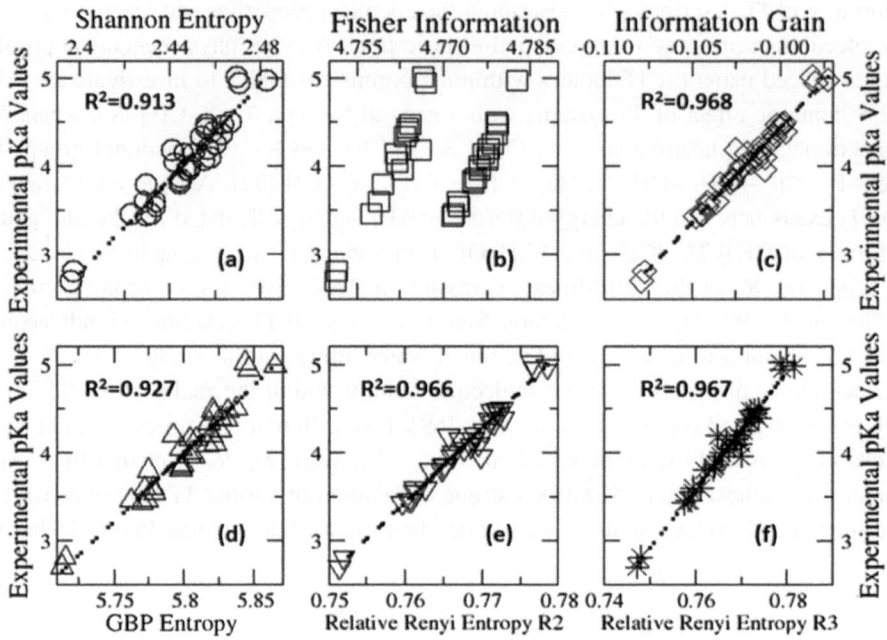

Fig. 5 Plots of experimental pKa versus different IT quantities, viz., **a** S_S, **b** I_F, **c** I_G, **d** S_{GBP}, **e** R_2, **f** R_3. (Reprinted from Reference 69 with permission from John Wiley and Sons. © 2017 Wiley Periodicals, Inc.)

lines that denote two types of substitution on benzoic acid. The fitting parameters indicate that the aforementioned IT quantities are suitable to quantitatively measure different aspects of molecular acidity.

Very recently in 2021, He et al. [71] analyzed the conformational changes of two opposite benzene rings in porphyrinoid system and predicted the aromaticity within the framework of ITA. Singlet and triplet states with 0 and $+2$ charges of different conformations, viz., Mobius, Huckel, and twisted porphyrinoid systems are considered in this study. The four model systems are considered i.e., Singlet 28, Singlet 30, Triplet 28, and Triplet 30. Figure 6 shows a strong correlation between the aromatic property (NICS) of the aforementioned system with the IT descriptor, I_G. An opposite correlation between NICS and I_G is unveiled when different conformation changes proceed.

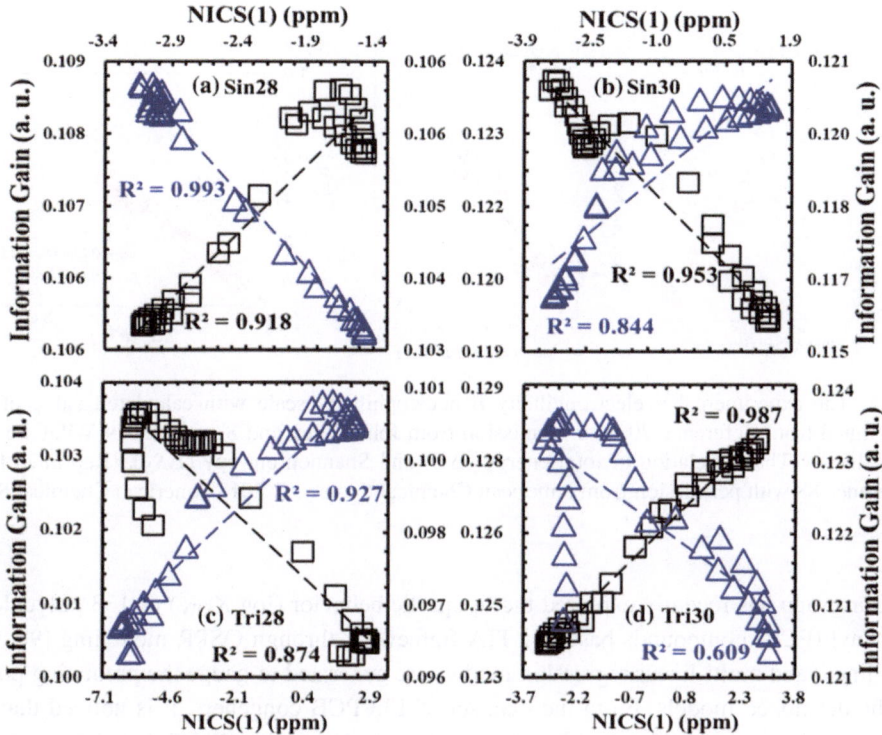

Fig. 6 NICS versus Information Gain from rotations of the two benzene rings. (Reprinted from Reference 71 with permission from Springer Nature. © 2021, the author(s), under exclusive licence to Springer-Verlag GmbH Germany, part of Springer Nature)

Various applications of IT descriptors in many chemical problems such as stability, electrophilicity, nucleophilicity, covalent and non-covalent interactions are also reported [70]. Figures 7a and b show strong correlations of experimental electrophilicity and nucleophilicity with the I_G descriptor [70]. The isomeric stability of C_{44}, C_{48}, C_{52}, and C_{60} fullerene systems and the correlation between their total energy difference with the difference in Shannon entropy is also mentioned (Fig. 7c) [88]. ITA quantities are also able to quantify the orbital-based properties for five polymeric systems such as repeating units of (1) trans-ethylene, (2) benzene, (3) naphthalene, (4) di-thiophene, and (5) phenanthrene chains [89].

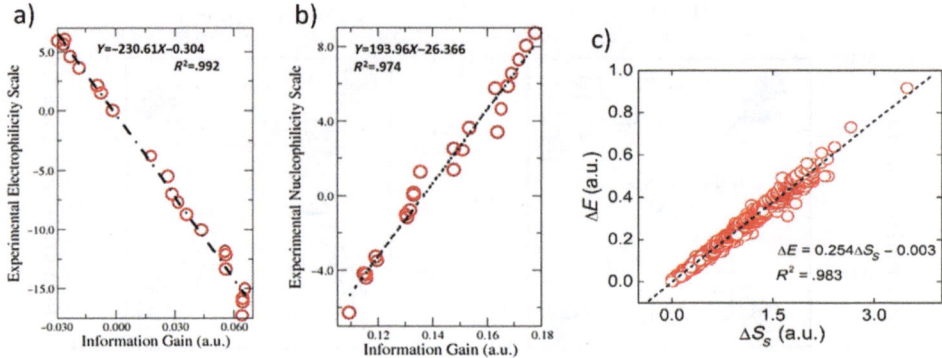

Fig. 7 The experimental **a** electrophilicity **b** nucleophilicity scale with calculated value of I_G. (Reprinted from Reference 70 with permission from John Wiley and Sons. © 2019 Wiley Periodicals, Inc.) **c** The correlation of total energy (ΔE) and Shannon entropy (ΔS_S). (Reprinted from Reference 88 with permission from American Chemical Society. © 2018 American Chemical Society)

Our group has recently explored the lipophilic behavior (log K_{OW}) of 133 polychlorobiphenyl (PCB) compounds based on ITA framework through QSPR modelling [90, 91]. The linear and multi-linear regression models are generated to gauge the predicting power of the developed models. From the data set of 133 PCB congeners, it is noticed that the linear combination of IT quantities (S_S and S_{GBP}) along with CDFT descriptors (ω and ω^2) can provide a reasonably good coefficient of determination value. Figure 8 shows the fitting of the calculated log K_{OW} versus experimental log K_{OW} for the test set constructed by using the training set model. The plots exhibit a good correlation between experimental log K_{OW} and calculated log K_{OW} values for almost all the cases. Using these descriptors together generates adequate descriptions for the calculated log K_{OW} of selected PCBs. Our analysis showed that the IT and CDFT descriptors used here are statistically significant in predicting the log K_{OW} value of the PCB congeners considered in this study.

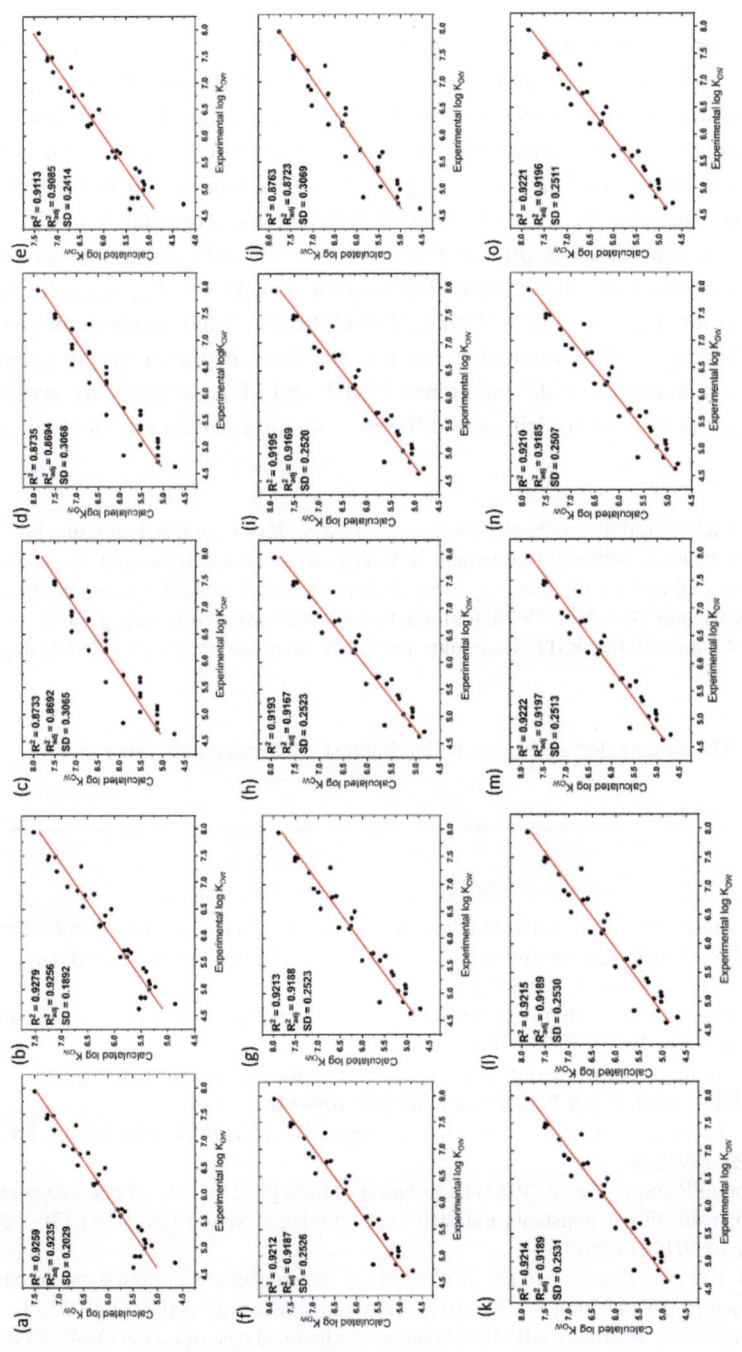

Fig. 8 Experimental and calculated values of log K_{OW} with **a** ω, **b** ω^2, **c** S_S, **d** S_{GBP}, **e** ω, ω^2, **f** ω, S_S, **g** ω, S_{GBP}, **h** ω^2, S_S, **i** ω^2, S_{GBP}, **j** S_S, S_{GBP}, **k** ω, ω^2, S_S, $1\,\omega$, ω^2, S_{GBP}, **m** ω, S_S, S_{GBP}, **n** ω^2, S_S, S_{GBP}, **o** ω, ω^2, S_S, S_{GBP} as descriptors for the model constructed using the test set of 33 PCBs. (Reprinted from Reference 90 with permission from Springer Nature. © 2023, the author(s), under exclusive licence to Springer Nature Switzerland AG)

4   Conclusion

This chapter focuses on the development of robust QSAR models for explaining toxicity as well as various structural parameters and properties using CDFT and IT descriptors. Linear and multi-linear regression models are constructed using some electronic descriptors, viz., electrophilicity (ω), its square term (ω^2), its cubic term (ω^3), local philicity (ω_m^+/ω_m^-), group philicities (ω_g^+ and ω_g^-), log P, (log $P)^2$, number of atoms in the molecule, etc. as the independent variables against different experimental toxicities as dependent variables through CDFT approach. Furthermore, different structural properties of chemical systems are discussed using some IT descriptors i.e., $S_S, I_F, E_n, S_{GBP}, I_G, R_n$. The lipophilic behaviour (log K_{OW}) of 133 polychlorobiphenyl (PCB) compounds based on ITA framework through QSPR modelling has recently been explored by our group. The overall discussion showed that the mentioned CDFT and IT descriptors are statistically significant in predicting the toxicity and different structural parameters of chemical systems.

Acknowledgements PKC would like to thank Professor Subhash C. Basak for kindly inviting him to contribute a chapter in the book entitled "Mathematical descriptors of molecules and biomolecules: Development and their applications in chemistry, drug design, chemical toxicology, and computational biology". He also thanks DST, New Delhi, for the J. C. Bose National Fellowship, grant number SR/S2/JCB-09/2009. AP and RP thank IIT Kharagpur and CSIR, respectively, for their fellowships.

Conflict of Interest The authors declare no competing interest, financial and/or otherwise.

References

1. Crum- Brown A, Fraser TR (1868–1869) On the connection between chemical constitution and physiological action. Part II—On the physiological action of the ammonium bases derived from Atropia and Conia. Trans R Soc 25: 693–739
2. Richet C (1893) On the relationship between the toxicity and the physical properties of substances. Compt Rend Soc Biol 9(5):775–776
3. Meyer H (1899) On the theory of alcohol narcosis I. Which property of anesthetics gives them their narcotic activity? Arch Exper Pathol Pharmakol 42:109–118
4. Overton E (1897) Osmotic properties of cells in the bearing on toxicology and pharmacology. Z Physik Chem 22:189–209
5. Hansch C, Maloney PP, Fujita T et al (1962) Correlation of biological activity of phenoxyacetic acids with Hammett substituent constants and partition coefficients. Nature 194(4824):178–180. https://doi.org/10.1038/194178b0
6. Hansch C, Fujita T (1964) P-σ-π analysis. A method for the correlation of biological activity and chemical structure. J Am Chem Soc 86:1616–1626. https://doi.org/10.1021/ja01062a035
7. Karelson M, Lobanov VS, Katritzky AR (1996) Quantum-chemical descriptors in QSAR/QSPR studies. Chem rev 96(3):1027–1044. https://doi.org/10.1021/cr950202r

8. Russom CL, Bradbury SP, Broderius SJ et al (1997) Predicting modes of toxic action from chemical structure: acute toxicity in the fathead minnow (Pimephales promelas). Environ Toxicol Chem 16(5):948–967. https://doi.org/10.1002/etc.5620160514

9. Kim KH (1993) 3D-quantitative structureactivity relationships: describing hydrophobic interactions directly from 3D structures using a comparative molecular field analysis (CoMFA) approach. Quant Struct-Act Relat 12(3):232–238. https://doi.org/10.1002/qsar.19930120303

10. Zhao YH, Ji GD, Cronin MTD et al (1998) QSAR study of the toxicity of benzoic acids to Vibrio fischeri, Daphnia magna and carp. Sci Total Environ 216(3):205–215. https://doi.org/10.1016/S0048-9697(98)00157-0

11. Raevsky O, Skvortsov V (2005) Quantifying hydrogen bonding in QSAR and molecular modeling. SAR QSAR Environ Res 16(3):287–300. https://doi.org/10.1080/10659360500036893

12. Parthasarathi R, Padmanabhan J, Subramanian V et al (2003) Chemical reactivity profiles of two selected polychlorinated biphenyls. J Phys Chem A 107(48):10346–10352. https://doi.org/10.1021/jp035620b

13. Parthasarathi R, Subramanian V, Roy DR et al (2004) Electrophilicity index as a possible descriptor of biological activity. Bioorg Med Chem 12(21):5533–5543. https://doi.org/10.1016/j.bmc.2004.08.013

14. Roy DR, Pal N, Mitra A et al (2007) An atom counting strategy towards analyzing the biological activity of sex hormones. Eur J Med Chem 42:1365–1369. https://doi.org/10.1016/j.ejmech.2007.01.028

15. Chakraborty A, Pan S, Chattaraj PK (2013) Biological activity and toxicity: a conceptual DFT approach. In: Applications of density functional theory to biological and bioinorganic chemistry, pp 143–179

16. Giri S, Chakraborty A, Gupta AK et al (2012) Modeling ecotoxicity as applied to some selected aromatic compounds: a conceptual DFT based quantitative-structure-toxicity-relationship (QSTR) analysis. In: Advanced methods and applications in chemoinformatics: research progress and new applications. IGI Global, pp 1–24

17. Pan S, Gupta A, Roy DR et al (2016) Application of conceptual density functional theory in developing QSAR models and their usefulness in the prediction of biological activity and toxicity of molecules. Chemometrics applications and research. Apple Academic Press, New York, pp 211–242

18. Pan S, Gupta AK, Subramanian V, Chattaraj PK (2017) Quantitative structure-activity/property/toxicity relationships through conceptual density functional theory-based reactivity descriptors. In: Pharmaceutical sciences. IGI Global, pp 1517–1572. https://doi.org/10.4018/978-1-5225-1762-7.ch058

19. Nantasenamat C, Isarankura-Na-Ayudhya C, Naenna T et al (2009) A practical overview of quantitative structure-activity relationship. Excli J 8:74–88

20. Guha R, Jurs PC (2004) Development of linear, ensemble, and nonlinear models for the prediction and interpretation of the biological activity of a set of PDGFR inhibitors. J Chem Inf Comput Sci 44(6):2179–2189. https://doi.org/10.1021/ci049849f

21. Hemmateeneja B, Safarpour MA, Miri R et al (2005) Toward an optimal procedure for PC-ANN model building: prediction of the carcinogenic activity of a large set of drugs. J Chem Inf Model 45(1):190–199. https://doi.org/10.1021/ci049766z

22. Baurin N, Mozziconacci JC, Arnoult E, Chavatte P, Marot C, Morin-Allory L (2004) 2D QSAR consensus prediction for high-throughput virtual screening. An application to COX-2 inhibition modeling and screening of the NCI database. J Chem Info Comput Sci 44(1):276–285. https://doi.org/10.1021/ci0341565

23. Itskowitz P, Tropsha A (2005) K nearest neighbors QSAR modeling as a variational problem: theory and applications. J Chem Inf Model 45(3):777–785. https://doi.org/10.1021/ci049628+

24. Dudek AZ, Arodz T, Gálvez J (2006) Computational methods in developing quantitative structure-activity relationships (QSAR): a review. Comb Chem High Throughput Screen 9(3):213–228. https://doi.org/10.2174/138620706776055539

25. Jana G, Pal R, Sural S, Chattaraj PK (2020) Quantitative structure-toxicity relationship: an "in silico study" using electrophilicity and hydrophobicity as descriptors. Int J Quantum Chem 120(6):e26097. https://doi.org/10.1002/qua.26097

26. Pal R, Jana G, Sural S, Chattaraj PK (2019) Hydrophobicity versus electrophilicity: a new protocol toward quantitative structure–toxicity relationship. Chem Biol Drug Des 93(6):1083–1095. https://doi.org/10.1111/cbdd.13428

27. Devillers J (ed) (1996) Network in QSAR and QSPR. Neural networks in QSAR and drug design, Academic Press, p 1

28. Calais JL (1993) Density-functional theory of atoms and molecules. In: Parr RG, Yang W (eds) Oxford University Press, New York, Oxford, 1989. Int J Quantum Chem 47:101. https://doi.org/10.1002/qua.560470107

29. Chermette H (1999) Chemical reactivity indexes in density functional theory. J Comput Chem 20:129–154. https://doi.org/10.1002/(SICI)1096-987X(19990115)20:1%3c129::AID-JCC13%3e3.0.CO;2-A

30. Geerlings P, De Proft F, Langenaeker W (2003) Conceptual density functional theory. Chem Rev 103:1793–1874. https://doi.org/10.1021/cr990029p(b)ChattarajPK(ed)(2005)Specialissue onchemicalreactivity.JChemSci117

31. Chattaraj PK, Nath S, Maiti B (2003) Reactivity descriptors. In: Tollenaere J, Bultinck P, Winter HD, Langenaeker W (eds) Computational medicinal chemistry for drug discovery, Chapter 11, Marcel Dekker, New York, pp 295–322

32. Parthasarathi R, Padmanabhan J, Subramanian V, Sarkar U, Maiti B, Chattaraj PK (2003) Toxicity analysis of benzidine through chemical reactivity and selectivity profiles: a DFT approach. Internet Electron J Mol Des 2:798–813

33. Parthasarathi R, Padmanabhan J, Subramanian V, Maiti B, Chattaraj, PK (2004) Toxicity analysis of 33'44'5-pentachloro biphenyl through chemical reactivity and selectivity profiles. Curr Sci 86: 535–542. https://www.jstor.org/stable/24107906

34. Chattaraj PK, Roy D, Giri S et al (2007) An atom counting and electrophilicity based QSTR approach. J Chem Sci 119:475–488. https://doi.org/10.1007/s12039-007-0061-1

35. Chattaraj PK, Parr RG (1993) Density functional theory of chemical hardness. In: Sen KD (eds) Chemical hardness. Structure and bonding, vol 80. Springer, Berlin, Heidelberg. pp 11–25

36. Chattaraj PK, Poddar A, Maiti B (2002) Chemical reactivity and dynamics within a density-based quantum mechanical framework. Reviews of modern quantum chemistry: a celebration of the contributions of Robert G Parr, vol 2. World Scientific, River Edge, pp 871–935

37. Chattaraj PK (2009) Chemical reactivity theory: a density functional view. CRC Press, Boca Raton

38. Kohn W, Becke AD, Parr RG (1996) Density functional theory of electronic structure. J Phys Chem 100:12974–12980. https://doi.org/10.1021/jp9606691

39. Pauling L (3rd ed) (1960) The nature of the chemical bond. Cornell University Press, Ithaca, NY

40. Sen KD, Jorgenson CK (1987) Electronegativity: structure and bonding, vol 66. Springer, Berlin

41. Parr RG, Donnelly RA, Levy M et al (1978) Electronegativity: the density functional viewpoint. J Chem Phys 68:3801. https://doi.org/10.1063/1.436185

42. Parr RG, Yang W (1989) Density functional theory of atoms and molecules. Oxford University Press, Oxford, U.K

43. Yang W, Mortier WJ (1986) The use of global and local molecular parameters for the analysis of the gas-phase basicity of amines. J Am Chem Soc 108:5708–5711. https://doi.org/10.1021/ja00279a008

44. Yang WT, Parr RG (1985) Hardness, softness, and the Fukui function in the electronic theory of metals and catalysis. Proc Natl Acad Sci USA 82:6723–6726. https://doi.org/10.1073/pnas.82.20.6723

45. Pearson RG (1997) Chemical hardness: applications from molecules to solids. Wiley-VCH, Weinheim

46. Parr RG, Szentpaly LV, Liu S (1999) Electrophilicity index. J Am Chem Soc 121:1922–1924. https://doi.org/10.1021/ja983494x

47. Mulliken RS (1955) Electronic population analysis on LCAO–MO molecular wave functions I. J Chem Phys 23:1833–1840. https://doi.org/10.1063/1.1740588

48. Parr RG, Yang W (1984) Density functional approach to the frontier-electron theory of chemical reactivity. J Am Chem Soc 106:4049–4050. https://doi.org/10.1021/ja00326a036

49. Yang W, Mortier WJ (1986) The use of global and local molecular parameters for the analysis of the gas-phase basicity of amines. J Am Chem Soc 108(19):5708–5711. https://doi.org/10.1021/ja00279a008

50. Chattaraj PK, Maiti B, Sarkar U (2003) Philicity: a unified treatment of chemical reactivity and selectivity. J Phys Chem A 107:4973–4975. https://doi.org/10.1021/jp034707u

51. Parr RG, Chattaraj PK (1991) Principle of maximum hardness. J Am Chem Soc 113:1854–1855. https://doi.org/10.1021/ja00005a072

52. Chamorro E, Chattaraj PK, Fuentealba P (2003) Variation of the electrophilicity index along the reaction path. J Phys Chem A 107:7068–7072. https://doi.org/10.1021/jp035435y

53. Parthasarathi R, Elango M, Subramanian V, Chattaraj PK (2005) Variation of electrophilicity during molecular vibrations and internal rotations. Theor Chem Acc 113:257–266. https://doi.org/10.1007/s00214-005-0634-3

54. Noorizadeh S (2007) Is there a minimum electrophilicity principle in chemical reactions? Chin J Chem 25:1439–1444. https://doi.org/10.1002/cjoc.200790266

55. Chattaraj PK, Sengupta S (1996) Popular electronic structure principles in a dynamical context. J Phys Chem 100:16126–16130. https://doi.org/10.1021/jp961096f

56. Liu SB (2009) Conceptual density functional theory and some recent developments. Acta Phys Chim Sin 25(3):590–600. https://doi.org/10.3866/PKU.WHXB20090332

57. Chattaraj PK, Chamorro E, Fuentealba P (1999) Chemical bonding and reactivity: a local thermodynamic viewpoint. Chem Phys Lett 314:114–121. https://doi.org/10.1016/S0009-2614(99)01114-8

58. Nalewajski RF, Parr RG (2000) Information theory, atoms in molecules, and molecular similarity. Proc Natl Acad Sci 97:8879–8882. https://doi.org/10.1073/pnas.97.16.8879

59. Nalewajski RF, Parr RG (2001) Information theory thermodynamics of molecules and their Hirshfeld fragments. J Phys Chem A 105:7391–7400. https://doi.org/10.1021/jp004414q

60. Nalewajski RF, Witka E, Michalak A (2002) Information distance analysis of molecular electron densities. Int J Quantum Chem 87:198–213. https://doi.org/10.1002/qua.10100

61. Nalewajski RF (2003) Information principles in the theory of electronic structure. Chem Phys Lett 372:28–34. https://doi.org/10.1016/S0009-2614(03)00335-X

62. Ayers PW (2006) Information theory, the shape function, and the hirshfeld atom. Theor Chem Acc 115:370–378. https://doi.org/10.1007/s00214-006-0121-5

63. Borgoo A, Geerlings P, Sen KD (2008) Electron density and Fisher information of Dirac-Fock atoms. Phys Lett Sect A Gen At Solid State Phys 372:5106–5109. https://doi.org/10.1016/j.physleta.2008.05.072

64. Geerlings P, Borgoo A (2011) Information carriers and (reading them through) information theory in quantum chemistry. Phys Chem Chem Phys 13:911–922. https://doi.org/10.1039/c0cp01046d

65. Alipour M (2013) Wave vector, local momentum and local coordinate from the perspective of information theory. Mol Phys 111:3246–3248. https://doi.org/10.1080/00268976.2013.777814

66. Alipour M (2015) Making a happy match between orbital-free density functional theory and information energy density. Chem Phys Lett 635:210–212. https://doi.org/10.1016/j.cplett.2015.06.073

67. Xu JH, Guo LY, Su HF et al (2017) Heptanuclear CoII5CoIII2 cluster as efficient water oxidation catalyst. Inorg Chem 56:1591–1598. https://doi.org/10.1021/acs.inorgchem.6b02698

68. Chen J, Liu S, Li M et al (2020) A density functional theory and information-theoretic approach study of chiral molecules in external electric fields. Chem Phys Lett 757:137858. https://doi.org/10.1016/j.cplett.2020.137858

69. Cao X, Rong C, Zhong A et al (2018) Molecular acidity: An accurate description with information-theoretic approach in density functional reactivity theory. J Comput Chem 39:117–129. https://doi.org/10.1002/jcc.25090

70. Rong C, Wang B, Zhao D, Liu S (2020) Information-theoretic approach in density functional theory and its recent applications to chemical problems. Wiley Interdiscip Rev Comput Mol Sci 10:1–22. https://doi.org/10.1002/wcms.1461

71. He X, Li M, Yu D et al (2021) Conformational changes for porphyrinoid derivatives: an information-theoretic approach study. Theor Chem Acc 140:1–8. https://doi.org/10.1007/s00214-021-02824-y

72. Cao X, Liu S, Rong C et al (2017) Is there a generalized anomeric effect? Analyses from energy components and information-theoretic quantities from density functional reactivity theory. Chem Phys Lett 687:131–137. https://doi.org/10.1016/j.cplett.2017.09.017

73. Flores-Gallegos N (2016) Informational energy as a measure of electron correlation. Chem Phys Lett 666:62–67. https://doi.org/10.1016/j.cplett.2016.10.075

74. Flores-Gallegos N (2018) Tsallis' entropy as a possible measure of the electron correlation in atomic systems. Chem Phys Lett 692:61–68. https://doi.org/10.1016/j.cplett.2017.12.014

75. Fisher RA (1925) Theory of statistical estimation. Math Proc Cambridge Philos Soc 22:700–725. https://doi.org/10.1017/S0305004100009580

76. Shannon CE (1948) A mathematical theory of communication. Bell Syst Tech J 27:379–423. https://doi.org/10.1002/j.1538-7305.1948.tb01338.x

77. Onicescu O (1966) Théorie de l'information. Energie informationnelle. CR Acad Sci Paris 263A:841–842

78. Ghosh SK, Berkowitz M, Parr RG (1984) Transcription of ground-state density-functional theory into a local thermodynamics. Proc Natl Acad Sci 81:8028–8031. https://doi.org/10.1073/pnas.81.24.8028

79. Alipour M, Badooei Z (2018) Toward electron correlation and electronic properties from the perspective of information functional theory. J Phys Chem A 122:6424–6437. https://doi.org/10.1021/acs.jpca.8b05703

80. Chattaraj PK, Chakraborty A, Giri S (2009) Net electrophilicity. J Phys Chem A 113(37):10068–10074. https://doi.org/10.1021/jp904674x

81. Renyi A (1970) Probability theory. North-Holland Publishing Company, Amsterdam

82. Tsallis C (1988) Possible generalization of Boltzmann-Gibbs statistics. J Stat phys 52(1):479–487. https://doi.org/10.1007/BF01016429

83. Roy DR, Parthasarathi R, Maiti B et al (2005) Electrophilicity as a possible descriptor for toxicity prediction. Bioorg Med Chem 13:3405–3412. https://doi.org/10.1016/j.bmc.2005.03.011

84. Hawkins DM, Basak SC, Mills D (2003) Assessing model fit by cross-validation. J Chem Inf Comput Sci 43:579–586. https://doi.org/10.1021/ci025626i
85. Kraker JJ, Hawkins DM, Basak SC, Natarajan R, Mills D (2007) Quantitative structure-activity relationship (QSAR) modeling of juvenile hormone activity: comparison of validation procedures. Chemometr Intell Lab Syst 87:33–42. https://doi.org/10.1016/j.chemolab.2006.03.001
86. Padmanabhan J, Parthasarathi R, Subramanian V, Chattaraj PK (2006) Group philicity and electrophilicity as possible descriptors for modeling ecotoxicity applied to chlorophenols. Chem Res Toxicol 19:356–364. https://doi.org/10.1021/tx050322m
87. Pal R, Pal G, Jana G, Chattaraj PK (2019) An in silico QSAR model study using electrophilicity as a possible descriptor against T. Brucei. Int J Chemoinformatics Chem Eng 8:57–68. https://doi.org/10.4018/IJCCE.20190701.oa1
88. Zhao D, Liu S, Rong C et al (2018) Toward understanding the isomeric stability of fullerenes with density functional theory and the information-theoretic approach. ACS Omega 3(12):17986–17990. https://doi.org/10.1021/acsomega.8b02702
89. Huang Y, Rong C, Zhang R et al (2017) Evaluating frontier orbital energy and HOMO/LUMO gap with descriptors from density functional reactivity theory. J Mol Model 23(1):1–12. https://doi.org/10.1007/s00894-016-3175-x
90. Poddar A, Pal R, Rong C et al (2023) A conceptual DFT and information-theoretic approach towards QSPR modeling in polychlorobiphenyls. J Math Chem 61:1143–1164. https://doi.org/10.1007/s10910-023-01457-9
91. Poddar A, Chordia A, Chattaraj PK (2024) QSPR models for n-octanol/water partition coefficient and enthalpy of vaporization using CDFT and information theory-based descriptors. J Chem Sci 136(2):23. https://doi.org/10.1007/s12039-024-02250-0

Complexity of Molecular Ensembles with Basak's Indices: Applying Structural Information Content

Denis Sabirov, Alexandra Zimina, and Igor Shepelevich

Abstract

Information-entropy-based structural descriptors are widely used for quantifying complexity of molecular objects in mathematical chemistry and chemometrics. Their application was exhaustively elaborated to isolated molecules—not molecular ensembles (collectives consisting of two or more molecules). Previously, we have deduced and justified the formula for calculating information entropy of molecular ensemble. This formula operates with information entropy in the original (Shannon) form. In this Chapter, we demonstrate the application of our approach to structural information content, which is one of the information-entropy-based descriptors introduced to theoretical chemistry by Prof. S. Basak's group. We discuss the properties of this combination. As an advantage, the approach allows separate ranking the contributions to the complexity of the molecular ensemble, associated with the molecular size and molecular structure, using a universal scale from 0 to 1.

Keywords

Information entropy · Structural information content · Cooperative entropy · Molecular ensemble · Molecular size · Molecular structure

D. Sabirov (✉) · A. Zimina · I. Shepelevich
Institute of Petrochemistry and Catalysis, UFRC RAS, Ufa 450075, Russia
e-mail: diozno@mail.ru

© The Author(s), under exclusive license to Springer Nature Switzerland AG 2025   113
S. C. Basak (ed.), *Mathematical Descriptors of Molecules and Biomolecules*, Synthesis
Lectures on Mathematics & Statistics, https://doi.org/10.1007/978-3-031-67841-7_6

1   Introduction

Information entropy became usual quantity in chemistry and one of its efficient applications relates to assessing the complexity of molecular graphs [1–10]. For this purpose, the information entropy is used in original form [11]:

$$h = -\sum_{j=1}^{n} p_i \log_2 p_i \tag{1}$$

where p_i are the probabilities of finding the same structural elements in the molecule (equivalent atoms, chemical bonds, or molecular moieties). In this form, information-entropy approach is still applied to chemical problems as a descriptor of molecular complexity. These applications must account that information entropy was introduced to the quantitative theory of information processing and, hence, it ignores semantic features [12]. It means that information entropy is better applied to the series of related molecules (isomers, homologs, molecules with similar structural patterns). In other words, when we narrow the correlation field, we select one semantic field and, therefore, overcome the mentioned disadvantage of h. If the series of the molecules are appropriately chosen, the h values catch most structural features such as elemental diversity and symmetry. In our studies, we partition the molecule over atom types, i.e. Equation (1) operates with the cardinalities (p_i) of the subsets representing the groups of chemically equivalent atoms. We previously used information entropy for sorting fullerene isomers, oxygen allotropes, interstellar molecules, isentropic species etc. [13–16].

When studying multimolecular objects, we found the non-additive behavior of information entropy [17, 18]. In other words, the information entropy of the molecular ensemble (h_{ME}) is not a sum of the h_i values of its constituent molecules:

$$h_{ME} = H_\Omega + \sum_{i=1}^{m} \omega_i h_i \tag{2}$$

Here the ensemble consists of the molecules of m types and ωi are their fractions:

$$\omega_i = \frac{N_i}{\sum_{i=1}^{m} N_i} \tag{3}$$

where, N_i are the numbers of atoms in i-th molecule. The first term of Eq. (2), called the cooperative entropy, emerges when the molecules are mixed to form the ensemble. Herewith, we follow the definition of molecular ensemble by [19]: molecular ensemble is just a set of the molecules, computationally treated together (and this concept should not be confused with molecular complex: molecular complex represents a novel chemical entity whereas molecular ensemble does not).

We discussed the meaning of the cooperative entropy in review [9]. In brief, it is crucial to make the numerical estimates consistent with chemical sense of molecular complexity. This parameter depends only on the partition of the molecular ensemble over its molecules and is defined as:

$$H_\Omega = -\sum_{i=1}^{m} \omega_i \log_2 \omega_i \qquad (4)$$

The maximal cooperative entropy is $H_\Omega^{max} = 1$. This value is achieved, for example, for the ensembles, made up with different atoms (when $\omega_i = 1/m$), or in the case of bimolecular ensembles, composed of the different molecules with the same size ($\omega_i = \frac{1}{2}$). Even in the case of the ensembles of m identical molecules with information entropy h, we have equal ω_i values and $H_\Omega \neq 0$, so that and $h_{ME} \neq h$.

Note that the h_i values depend on the molecular structure and independent of the molecular size [14]; in contrast ω_i and H_Ω depend only on the molecules' size. This allows us concluding that there are two types of terms in Eq. (2), viz. depending only on the molecular size (H_Ω) and depending on both the size and the structure ($\Sigma \ \omega_i h_i$). Such a 'size—structure' discrimination becomes useful for the information-entropy-based approach to quantifying the complexity changes under chemical reactions [20].

Prof. S. Basak's group works on structural descriptors and then widely uses them for quantitative structure–activity relationship (QSAR) and other chemometric studies [21–26]. This group started using information entropy, calculated according Eq. (1) under the name 'information content' ($IC = h$) [22]. In the referenced work, other related descriptors were introduced. One of them is structural information content. It is placed in the title of this chapter due to its usefulness for describing the complexity of molecular ensembles in line with the approach we develop.

2 Structural Information Content of the Molecule

Structural information content (SIC) is introduced as an information-entropy based descriptor according to the formula below [22]:

$$SIC = \frac{h^{(N)}}{h_{max}^{(N)}} = \frac{h^{(N)}}{\log_2 N} \qquad (5)$$

where $h^{(N)}$ denotes the information entropy of the considered N-atomic molecule (or its information content in terms of the Basak approach) and $h_{max}^{(N)} = \log_2 N$ is the maximal information entropy achievable for the same number of atoms. The importance of treating the same number of atoms in the numerator and denominator of Eq. (5) is stressed by the upper indices of the used h values.

This quantity could be considered as a degree of structural inhomogeneity of the molecule in addition to h. However, the h (or IC) values are expressed in bits whereas SIC is dimensionless indicating the relation of the structural diversification of the molecule towards the maximally achievable one. Indeed, the SIC values lie in the range 0–1 regardless of the size of the molecule and equal 1 for the molecules with $h = h_{max}$.

Let us consider simple linear graphs (Fig. 1). They may correspond to carbon chains that approximate hydrocarbon molecules (alkanes) with omitted hydrogen atoms. The information entropy parameters of the graphs are shown in Table 1. This example demonstrates the order of calculating the SIC values.

As for the applications, SIC values are mostly used in QSAR models [22]. We previously used them as additional structural descriptors for discriminating interstellar molecules [13]. The mentioned studies stressed their better suitability for comparing molecules of different sizes than the original information entropy.

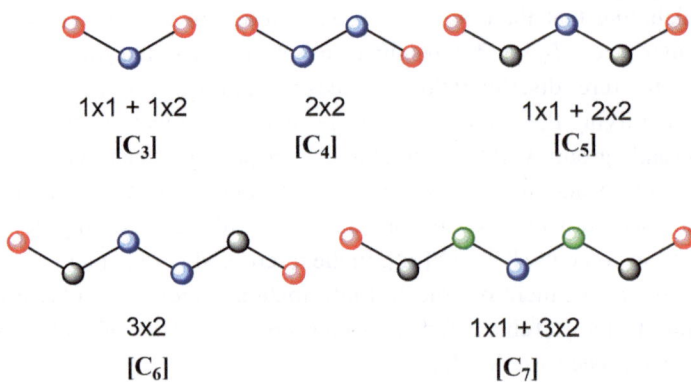

Fig. 1 Simple linear graphs [C_N] with colored inequivalent vertices and corresponding partition formulae ([number of atom types] × [number of atoms within the type])

Table 1 Information entropy parameters of linear chains

Chain [C_N]	Partition formula	h (bits)	h_{max} (bits)	SIC
[C_2]	1×2	0	1	0
[C_3]	$1 \times 1 + 1 \times 2$	0.918	1.585	0.579
[C_4]	2×2	1.000	2.000	0.500
[C_5]	$1 \times 1 + 2 \times 2$	1.522	2.322	0.655
[C_6]	3×2	1.950	2.585	0.613
[C_N]	$N/2 \times 2$ (even); $1 \times 1 + (N-1)/2 \times 2$ (odd)	$\log_2 \frac{N}{2}$ (even); $\frac{1}{N}\log_2 \frac{N}{2} + \frac{N-1}{N}\log_2 \frac{N}{2}$ (odd)	$\log_2 N$	h/h_{max}

3 Information Entropy Descriptors of Bimolecular Ensembles

3.1 Applying Structural Information Content to a Bimolecular Ensemble

Let us consider bimolecular ensemble A + B. Its information entropy is:

$$h_{ME}(A + B) = \omega_A h_A + \omega_B h_B + H_\Omega \tag{6}$$

$$H_\Omega = -\omega_A \log_2 \omega_A - \omega_B \log_2 \omega_B \tag{7}$$

To transit from the h (or IC) values, expressed in bits, to the dimensionless quantity of SIC, we divide each information-entropy term of Eq. (6) by the corresponding maximal value, similar to the case of introducing SIC of the molecule:

$$SIC_{ME}(A + B) = \omega_A \frac{h_A}{h_{max}^{(N_A)}} + \omega_B \frac{h_B}{h_{max}^{(N_B)}} + \frac{H_\Omega}{H_\Omega^{max}} \tag{8}$$

Taking into account that $h_{max}^{(N)} = \log_2 N$ and $H_\Omega^{max} = 1$, we obtain:

$$SIC_{ME}(A + B) = \omega_A SIC_A + \omega_B SIC_B + H_\Omega \tag{9}$$

As we mention above, the function ranges are following: $SIC \in [0; 1]$ and $H_\Omega \in [0; 1]$. Additionally, from the definition (Eq. (3)):

$$\omega_A + \omega_B = 1 \tag{10}$$

The above intervals define the range of dimensionless function SIC_{ME} as:

$$SIC_{ME} \in [0; 2] \tag{11}$$

Herewith, the ranges of the parts of SIC_{ME} are:

$$\omega_A SIC_A + \omega_B SIC_B \in [0; 1] \tag{12}$$

$$H_\Omega \in [0; 1] \tag{13}$$

Thus, using Eq. (9), we obtain two comparable scales from 0 to 1, in which two contributions to the SIC value of the ensemble are assessed. The first contribution $\omega_A SIC_A + \omega_B SIC_B$ is defined by molecular structure and molecular size, as it depends on SIC_i and ω_i, respectively. The second one (H_Ω) depends only on the size of molecules A and B.

3.2   Exemplification

We exemplify the application of SIC to bimolecular ensembles using the chains, described in Sect. 2. For this purpose, we consider the series of bimolecular ensembles, which contain two chains, constant [C_6] and variable [C_N]. We use below two modes of representing the results of the SIC calculations to demonstrate how they could be informatively visualized.

First, we use a histogram for this purpose. This mode allows monitoring the ratio of the contributions from $\Sigma\omega_i SIC_i$ and H_Ω to the structural information content of the molecular ensembles (Fig. 2). Interestingly, it shows that the size-dependent term (cooperative entropy H_Ω) dominates over the structure-and-size part $\Sigma\omega_i SIC_i$. The dependences of both contributions on the length of the variable chain are not monotonous. The $\Sigma\omega_i SIC_i$ values oscillate for even and odd N. This behavior is due to the different partition of even and odd chains over the atom types and it was previously discussed in details for the h values of alkane molecules [17]. Expectedly, more symmetric even chains [C_N] have lower h and SIC values as compared with their neighbors [$C_{N\pm1}$]. Dimensionless cooperative entropy, the second contribution to SIC_{ME}, has the maximum at $N = 6$. Its behavior is similar to its prototype, expressed in bits [27].

The histogram induced the following question. For small N it is true that $\Sigma\omega_i SIC_i < H_\Omega$. When the sign is reversed in this inequality? To answer, we have continued functions $\Sigma\omega_i SIC_i = f(N)$ and $H_\Omega = f(N)$ until the intersection point and represented these discrete functions as continual (Fig. 3). As follows from the plots, $\Sigma\omega_i SIC_i < H_\Omega$ up to $N = 22$, but for the next ensemble $\Sigma\omega_i SIC_i > H_\Omega$. We do not interpret this result in chemical terms and consider it only as a mathematical feature of the functions under discussion. Saw-like dependence $\Sigma\omega_i SIC_i = f(N)$ becomes smoothed when $N \rightarrow \infty$.

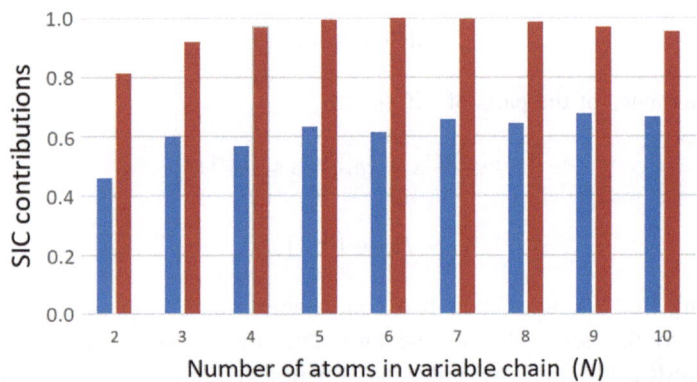

Fig. 2 The contributions to the SIC values of molecular ensembles [C_6] + [C_N] from $\Sigma\omega_i SIC_i$ (blue bars) and H_Ω (burgundy bars). Each contribution is dimensionless and varies within the range 0–1

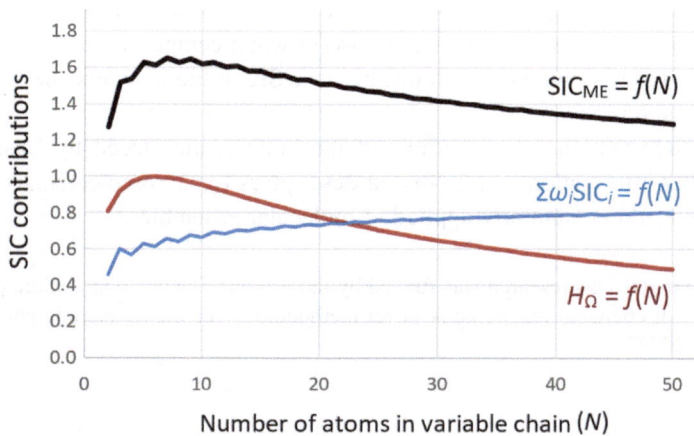

Fig. 3 Structural information contents of molecular ensembles $[C_6] + [C_N]$ and contributions $\Sigma\omega_i SIC_i$ and H_Ω

The maximal SIC_{ME} value, equal to 2, is never achieved for the considered molecular ensembles. The highest SIC_{ME} is 1.653 that corresponds to $N = 7$, i.e. the maximum of function $SIC_{ME} = f(N)$ is shifted relative to the maximum of $H_\Omega = f(N)$ ($N = 6$). This behavior is similar to the mismatch between the maximums of the original information entropy functions h_{ME} and H_Ω having dimensionality of bits [27].

Function $SIC_{ME} = f(N)$ inherits saw-like view of $\Sigma\omega_i SIC_i$. Thus, the structure of the chain (more general, chemical structure) affects the resulting SIC_{ME} values. Similar to function $\Sigma\omega_i SIC_i = f(N)$, the oscillations of SIC_{ME} are damped with $N \to \infty$. The discussed three functions have the following trends when $N \to \infty$. Cooperative entropy $H_\Omega \to 0$ according to Eq. (7) because the presence of chain $[C_6]$ in the ensemble $[C_6] + [C_N]$ becomes negligible, so that $\omega(C_6) \to 0$ and $\omega(C_N) \to 1$. Functions SIC_{ME} and $\Sigma\omega_i SIC_i$ approach to unity line from above and below, respectively. Just to demonstrate the slowness of this trend numerically: $SIC_{ME} \approx \Sigma\omega_i SIC_i = 0.957$ and $H_\Omega = 1.33 \cdot 10^{-5}$ for $N = 10^7$; $SIC_{ME} \approx \Sigma\omega_i SIC_i = 0.975$ and $H_\Omega = 2.32 \cdot 10^{-10}$ for $N = 10^{12}$.

4   Conclusion

We have demonstrated the suitability of structural information content as a structural descriptor for molecular ensembles. As applied to the ensembles, SIC_{ME} varies in the range 0–2 whereas both its contributions lie in the range 0–1. The contributions are associated with the complexity relating to molecular size (cooperative entropy H_Ω) or both molecular size and molecular structure ($\Sigma\omega_i SIC_i$). Thus, we obtain the tool for separate ranking these components of the complexity within one dimensionless scale.

We consider that this approach could be further used for comparative studies of various multimolecular ensembles. It is especially useful when comparing molecular ensembles, made up with different number of atoms, because SIC value is independent of molecular size from the definition.

Structural information content is one of the indices, introduced by Basak et al. We think that other information-entropy-based descriptors (e.g., complementary information content, CIC) could be similarly applied to molecular ensembles.

Acknowledgements This research was funded by the Russian Science Foundation, project "Information entropy of chemical reactions: A novel methodology for digital organic chemistry", grant number 22-13-20095.

References

1. Banaru A, Aksenov S, Krivovichev S (2021) Complexity parameters for molecular solids. Symmetry 13:1399
2. Barigye SJ, Marrero-Ponce Y, Pérez-Giménez F, Bonchev D (2014) Trends in information theory-based chemical structure codification. Mol Diversity 18:673–686
3. Basak SC, Gute BD, Grunwald GD (1997) Use of topostructural, topochemical, and geometric parameters in the prediction of vapor pressure: a hierarchical QSAR approach. J Chem Inf Comput Sci 37:651–655
4. Bonchev D (1995) Kolmogorov's information, Shannon's entropy, and topological complexity of molecules. Bul Chem Commun 28:567–582
5. Dehmer M, Mowshowitz A (2011) A history of graph entropy measures. Inf Sci 181:57–78
6. Ghorbani M, Rajabi-Parsa M, Majidi R, Mirzaie RA (2021) Novel results on entropy-based measures of fullerenes. Fullerenes, Nanotubes, Carbon Nanostruct 29:114–125
7. Krivovichev SV, Krivovichev VG, Hazen RM (2018) Structural and chemical complexity of minerals: correlations and time evolution. Eur J Mineral 30:231–236
8. Nalewajski R (2021) Resultant information descriptors, equilibrium states and ensemble entropy. Entropy 23:483
9. Sabirov DS, Shepelevich IS (2021) Information entropy in chemistry: an overview. Entropy 23:1240
10. Stankevich IM, Stankevich IV, Zefirov NS (1988) Topological indices in organic chemistry. Russ Chem Rev 57:191–208
11. Shannon CE (1948) A mathematical theory of communication. The Bell Syst Tech J 27:379–423
12. Kolchinsky A, Wolpert DH (2018) Semantic information, autonomous agency and non-equilibrium statistical physics. Interface Focus 8:20180041
13. Sabirov DS (2016) Information entropy of interstellar and circumstellar carbon-containing molecules: molecular size against structural complexity. Comput Theor Chem 1097:83–91
14. Sabirov D, Koledina K (2020) Classification of isentropic molecules in terms of Shannon entropy. EPJ Web Conf 244:01016
15. Sabirov DS, Ori O, Laszlo I (2018) Isomers of the C_{84} fullerene: a theoretical consideration within energetic, structural, and topological approaches. Fullerenes, Nanotubes, Carbon Nanostruct 26:100–110

16. Sabirov D, Shepelevich I (2015) Information entropy of oxygen allotropes. A still open discussion about the closed form of ozone. Comput Theor Chem 1073:61–66
17. Sabirov DS (2018) Information entropy changes in chemical reactions. Comput Theor Chem 1123:169–179
18. Sabirov DS (2020) Information entropy of mixing molecules and its application to molecular ensembles and chemical reactions. Comput Theor Chem 1187:112933
19. Ugi I, Gillespie P (1971) Representation of chemical systems and interconversions bybe matrices and their transformation properties. Angew Chem Int Ed 10:914–915
20. Sabirov DS, Tukhbatullina AA, Shepelevich IS (2022) Molecular size and molecular structure: discriminating their changes upon chemical reactions in terms of information entropy. J Mol Graph Model 110:108052
21. Basak SC, Grunwald GD (1994) Molecular similarity and risk assessment: analog selection and property estimation using graph invariants. SAR QSAR Environ Res 2:289–307
22. Basak SC, Harriss D, Magnuson V (1984) Comparative study of lipophilicity versus topological molecular descriptors in biological correlations. J Pharm Sci 73:429–437
23. Basak SC, Magnuson V, Niemi GJ, Regal RR (1988) Determining structural similarity of chemicals using graph-theoretic indices. Discr Appl Math 19:17–44
24. Basak SC, Majumdar S (2015) Current landscape of hierarchical QSAR modeling and its applications: some comments on the importance of mathematical descriptors as well as rigorous statistical methods of model building and validation. Adv Math Chem Appl 1:251–281
25. Basak SC, Mills DR (2001) Use of mathematical structural invariants in the development of QSPR models. Match 44:15–30
26. Basak SC, Niemi GJ, Veith GD (1990) Optimal characterization of structure for prediction of properties. J Math Chem 4:185–205
27. Tukhbatullina AA, Shayakhmetova RR, Sabirov DS (2022) Cooperative information entropy of a dimorphic molecular ensemble of open type. Vestnik Bashkirskogo Universiteta 27:352–354

Descriptors from Calculated Stereo-Electronic Properties and Molecular Electrostatic Potentials (MEPs) May Provide a Powerful "Interaction Pharmacophore" for Drug Discovery

Apurba K. Bhattacharjee

Abstract

Molecular Electrostatic Potentials (MEPs), their three dimensional profiles and descriptors are very useful representations for intrinsic reactivity of molecules. They enable to visualize charge distributions and charge related properties. MEP not only allows to visualize the size and shape but also provides an invaluable tool for predicting the behavior of complex molecules. Their profiles are particularly useful for medicinal and organic chemists. MEP being an experimentally determinable property contains a wealth of reliable information of intrinsic molecular and material interaction capabilities. It is therefore commonly known as "interaction pharmacophore". The present chapter first introduces the theoretical foundation of MEP and its mathematical expressions followed by various earlier application studies. Finally, the chapter describes MEP applications in drug design and discovery studies ranging from mutagenicity, carcinogenicity, to cardiotonics, anti-inflammatory, malarial, leishmanial, insect repellency, organo-phosphorous poisoning and antiviral compounds including identification of potential anti Covid-19 compounds from author's laboratory.

Keywords

Stereo-electronic properties • Molecular electrostatic potentials (MEPs) and profiles • Interaction pharmacophore models • Drug discovery and virtual screening

A. K. Bhattacharjee (✉)
Department of Microbiology and Immunology, School of Medicine, Biomedical Graduate Research Organization, Georgetown University, Washington, DC 20057, U.S.A.
e-mail: ab3094@georgetown.edu

1   Introduction

Amongst stereoelectronic properties, molecular electrostatic potential (MEP) is the most important and useful property in drug discovery, usually known as "interaction pharmacophore". An accurate estimation of interaction pharmacophore profile of a molecule can provide invaluable information for identifying new compounds.

The reactive nature of a molecule is critical for binding to the target structure and this information can be obtained from a MEP profile. Quantum chemical methods can be used for calculations of MEPs and corresponding profiles. The MEP profiles on the van der Waals surface quantifies the strength of nucleophilicity and electrophilicity, whereas profiles beyond the van der Waals surface are believed to be responsible for long-range recognition interactions between a molecule and its target receptor at larger distances of separation [1]. It is through this potential a molecule recognizes its receptor, and accordingly promotes interaction with complimentary sites of receptor. MEP quantifies the ability of nuclei and electrons of a molecule to create an electrostatic field in the surrounding space providing information regarding recognition interactions for noncovalent bonds at both near and long range distances. It can not only be calculated but also be experimentally determined by diffraction methods. However, ab intio quantum calculations at higher levels of theory on a molecular structure provide fairly accurate estimate of this property as well as provide a large well-defined information on its intrinsic reactivity. The information also include deeper insights of nonbonding long range recognition interactions, protein folding and pKa's of ionizable residues [2].

MEP is a physical property and the potential V(r) created in space around a molecule by its nuclei and electrons has been well documented as means for estimating electrophilicity, nucleophilicity and recognition interactions [3, 4]. It can be described as to how a molecule should first be encountered by another approaching molecule. The region of a molecule that has a negative electrostatic potential is susceptible to electrophilic attack whereas, the region that has a positive potential is susceptible for nucleophilic attack. Mathematically, MEP (r) is defined by Eq. (1).

The sign of MEP (r) at any point r^1 is the net result of the positive and negative contributions of the nuclei and electrons, where Z_α is the charge on nucleus α, located at R_α.

$$MEP(\mathrm{r}) = \sum_\alpha Z_\alpha / |r - R_\alpha| - \int_\rho dr^1 / |r - r^1| \qquad (1)$$

Since MEP profiles provide information about the charge distribution of a molecule, information on the charge distribution in a molecule including the property of the nucleus and nature of electrostatic potential energy can also be ascertained. Interactively, if a positive test probe is imagined to move along the spherical isosurface of an atom, the positively charged nucleus will experience a radially constant electric field. Thus, the region of higher average electrostatic potential energy would indicate the presence of a stronger positive charge or a weaker negative charge. Since the nuclei consistently carry of the positive charge, higher potential energy value would indicate fewer electrons in this region and thus, absence of negative charges. The converse is also true. This property of electrostatic holds to molecules as well. Mathematically the relationship between charge distribution and electrostatic potential may be represented by the following equation:

$$\text{Total Electrostatic Potential Energy} = \sum \text{Electrostatic Potential Energy}.$$

The equation is commonly used to find the electrostatic potential. The total energy of a pathway is the sum of the energies of the particle interacting with every electric field produced along the pathway.

Quantum chemical descriptors have been frequently used in QSAR and drug discovery studies because of a wealth of physical information embodied in many of these theoretical descriptors [5]. The ability of a bioactive molecule to interact with recognition sites in receptor proteins or active sites in enzymes results from a combination of steric and electronic properties. Therefore, the study of stereoelectronic properties in general and molecular electrostatic potentials in particular can provide a better understanding of mechanism of action and intrinsic "interaction pharmacophore" profile to aid design and search for more efficacious compounds. Since MEP profile is essentially the "interaction pharmacophore" of a molecule and therefore also provides necessary information for assessment of its electronic structure [6].

For analyzing an electrostatic potential map, it is important to consider the charge distribution. The relative distributions of electrons provide the crucial information of MEP maps and thus provide the relationship between electrostatic potential and charge distribution. A high (positive) electrostatic potential indicates the relative absence of electrons, usually represented by blue color and a low (negative) electrostatic potential indicates an abundance of electrons represented by red color. For example, in sulfur dioxide, oxygen has a greater electronegativity value than sulfur and therefore oxygen atoms would have a greater electron density around them than sulfur atoms. The electrostatic regions in molecules are usually represented as color-coded regions. Thus, for example MEP map of atropine on the van der Waals surface would look like (Fig. 1a) whereas for protonated atropine on the same van der Waals surface would look like Fig. 1b. Deepest red color in the map is the location of the most electron dense part of the molecule that is the nucleophilic site where as deepest blue color in the map would indicate the most electron deficient region and therefore the most electrophilic site.

MEP profiles beyond van der Waals surface are shown in (Fig. 1c–h) by taking the same example of atropine and its protonated form at different energies i.e., distances beyond van der Waals surface (Fig. 1). The MEP profiles beyond the van der Walls surface are very important for recognition of complementary sites of the target structure from a distance. The target structure recognizes these features at a distance and accordingly promotes complementary interactions. Profiles at −5.0 kcal/mol often show much

Fig. 1 MEP profiles at different distances with the atropine molecule: **a** left column: on the van der Waals surface of atropine, **b** right: on the van der Waals surface of protonated atropine; **c** left: at −20.0 kcal/mol of atropine, **d** right: at −20.0 kcal/mol protonated atropine; **e** left: at −10.0 kcal/mol atropine, **f** right: at −10.0 kcal/mol protonated atropine; **g** left: at −5.0 kcal/mol atropine **h** right: at −5.0 kcal/mol protonated atropine

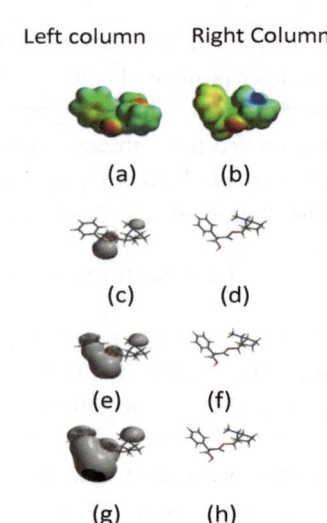

larger extended MEP profile with holes indicating a powerful nucleophilic suction-pump like appearance. This kind of profile of a ligand molecule can be rapidly recognized by the target protein for a rapid interaction prior to formation of any covalent bond. MEP interactions are primarily weaker in nature and mainly important for hydrogen bonds, hydrophobic interactions, pi-pi and cation-pi interactions. When atropine is protonated, the MEP profile beyond van der Waals structure does not appear because the proton is already bound to it. Molecules showing similar MEP profiles beyond van der Waals surface, particularly at −5.0 kcal/mol would indicate electrostatic bio-isosterism (similarity for recognition by the target structure) and therefore, likely to have similar interaction with the target protein resulting in similar bioactivities. Similarity searches with MEP profiles based on electrostatic bio-isosterism can be an useful approach for identification of similar bioactive compounds.

However, when a receptor (protein)—ligand (potential drug molecule) interaction takes place in solvent e.g., water, the hydrophobic complementarity is mainly related to hydration free energy as, ΔH (hydration)—$T\Delta S$ (hydration) where, free energy of host–guest interaction is represented as, ΔG (interaction) = δH (vacuum)—$T\delta S$ (vacuum)— [δH (solvent)—$T\delta S$ (solvent)].

It is well documented that host–guest interactions avoid steric conflicts but most of the docking procedures apply the principle of optimum filling of empty space and quite often recognition of crevice space in the host is important but ignored [7].

The present chapter would also attempt to provide an insight of how "interaction pharmacophore" models of compounds can be useful both qualitatively and quantitatively for structure–activity relationships in drug discovery. How from "interaction pharmacophore", statistical pharmacophore models can be generated and utilized for QSAR interpretations against biological activities. Which this can be useful for compound database searches to identify new potentially active compounds. The electrostatic bio-isosteric active groups observed in MEP profiles may serve as an important template for virtual searches due to similarity of recognition interactions with the receptor. Both literature based studies and studies performed in author's own lab are discussed below.

With the advent of virtual screening for bioactive compounds from databases, docking of compounds at the protein active site along with interaction pharmacophore modeling provided tremendous impetus to the current day drug discovery programs. However, for rapidly achieving the objective, a transition from quantum chemically developed "interaction pharmacophore" model to a statistically based pharmacophore model would be a more efficient approach. QM methods are usually more time consuming but accurate for developing MEP based interaction pharmacophore models. Therefore, important pharmacophore features, such as hydrogen acceptors, donors, hydrophobic regions obtained from QM based MEP profiles of a few active compounds can be utilized to develop statistically a training set of larger number active compounds from which a reliable pharmacophore model could be established. The model now would be far more efficient for search of new active compounds from databases. Since the foundation of the statistical model is fairly accurate, it would now be far more efficient and reliable not only to handle millions of compounds from databases at a shorter time but also be able to shortlist a handful new potentially active compounds for experimental testing.

In the context of potential drug molecules, it is important to understand how a molecule becomes a drug. A deeper look into potential drug molecules would reveal that these molecules are capable of optimally interacting with the active site of a disease specific receptor protein or enzyme to trigger or inhibit its biological response. The idea is similar to a lock and key mechanism. Now, at the molecular level, optimal interaction with the receptor would primarily be through a combination of steric and electronic features to trigger the biological response. Therefore, these features would be responsible for pharmacophore of activity of the drug molecule. The International Union of Pure and Applied Chemistry (IUPAC) in 1998 provided an official definition of pharmacophore [8]. It states that a pharmacophore is "an ensemble of steric and electronic features that are necessary for optimal interaction with a specific receptor target structure (a protein or an enzyme) to trigger or inhibit its biological response" [9]. However, to become a successful drug, the molecule will have to undergo several iterations of animal testing, toxicity evaluations and human clinical trials before being approved by the regulatory bodies for marketing in a pharmacy. Pharmacophores can provide useful templates for search of new compounds from databases, find out the active site of unknown target protein structures from binding features and can be insightful for designing novel ligands. As an example, it is shown in (Fig. 2a and b), how a statistical pharmacophore model was developed from a single molecule (anti-bacterial Thiolactomycin) by using MEP profile to a reasonable model for database searches to identify new compounds. The most negative potential site on the van der Waals surface is seen by the carbonyl oxygen atom of thiolactomycin which indicated the position to be most nucleophilic followed by another significant nucleophilic site by the hydroxyl moiety and rest of the molecule appeared hydrophobic due the large distribution of weak electrostatic potentials. These characteristics of thiolactomycin were further strengthened by profiles at 1.3–1.5 A away from van der Waals surface. Thus, the reasonable pharmacophore model created from these MEP features constituted two hydrogen bond acceptors and a hydrophobic feature (Fig. 2a). By mapping the features onto thiolactomycin and converting it to a shape based template for virtual screening of a in-house database resulted in the identification of a new anti-bacterial compound (Fig. 2b). Indeed the search led to identification of several potential anti-bacterial compounds [10].

Thus, the focus of the chapter will mostly be on ligand based drug design and discovery using interaction pharmacophores and statistical pharmacophores. Historically, quantum

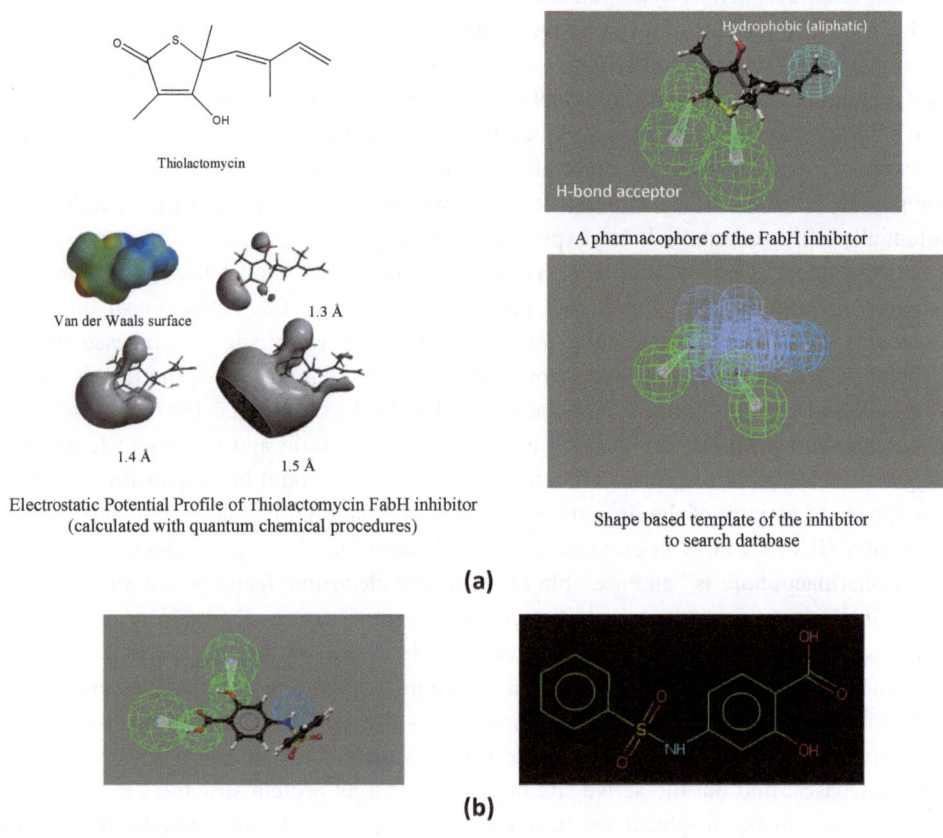

Fig. 2 a How a pharmacophore model was developed from a single molecule (anti-bacterial Thiolactomycin) and used for database searches to identify new compounds. Left column: structure of thiolactomycin and its MEP profiles. Right column: pharmacophore model from MEP profile and its shape based template for database search. **b** Structure of a newly identified antibacterial compound and perfect mapping of the pharmacophore on it

chemistry for small molecule ligands was used to investigate energy, geometry, and electronic features, such as orbitals, dipole moment, atomic charges, and electrostatics. Information obtained from the calculations were used for quantitative structure–activity relationship (QSAR) and 3D QSAR studies for better ligand design and optimization. However, our approach here was specifically on electrostatic potential descriptors and profiles of small bioactive molecules to find out how they can be related to biological activity and be useful for database searches for identifying potentially new active compounds. Earlier literature based studies and studies from author's laboratory are discussed.

2 Earlier Studies on Application of Molecular Electrostatic Potential (MEP) Descriptors

Over the past half a century, the molecular electrostatic potential (MEP) established itself to be an effective tool for providing important insights on molecular properties and interactions. With modern advances in computational efficiency, incredible memory, speed and technology, it can now be applied for obtaining accurate information of a variety of important chemical and biological processes. The range of its applicability is not only focused on sites for electrophilic and nucleophilic attack, but includes solvent effects, studies on macromolecules, molecular clusters, crystallographic studies, and both correlation and prediction of macroscopic properties. In addition, due to widespread use of DFT (density functional theory) instead of conventional HF (Hatree-Fock) quantum chemical calculations, molecular electrostatic potential is more often used for reliable determination of topological characteristics. This is because MEP can assess reasonably accurately the opposing contributions from nuclei and electrons of molecules [1].

One of the pioneers in the field of molecular electrostatic potential (MEP) studies on chemical reactivity is Tomasi et al. [11]. Tomasi and his coworkers first showed difficulties of MEP based interpretations of condensed systems as opposed to analogous isolated molecules. Tomasi et al. [11] showed that when chemical nature of molecular systems are combined with physical analysis of a molecular property, two distinct types of calculational procedures are necessary: (1) for low-polar solvents, since a weak H-bonding occurs between a solute and solvent (for example, acetonitrile in CHCl3), a statistical analysis of hybrid type of calculations would be a better option; and (2) for more polar solvents, where stronger H-bonding interactions between solute and solvent occurs (for example, diazines in water), a more accurate calculation would be required which can be obtained with smaller number of solvent molecules like molecular clusters. However, Tomasi et al. [11] recommended that for the second type of calculations (i) H-bonded structures of clusters are to be obtained using geometry optimization at high-level of QM techniques and (ii) the bulk effects are to be calculated through an accurate continuum model. The significance of the study is not only the use of molecular electrostatic potential as an interpretative tool for intermolecular interactions, but also the necessary use of HF level calculations for description of intermolecular energy, molecular electrostatics and semiclassical approximation for solvation effects.

Next, Leboeuf et al. [12] showed how MEP could be reasonably well calculated using density functional theory as simplified analytic expressions. The researchers presented critical points of MEPs and showed how MEP evolves in chemical reactions.

In another study [13], electrostatic potential fields for QSAR/QSPR (Quantitative Structure Activity/Property Relationships (QSAR/QSPR) were used for establishment and virtual high-throughput screening (VHTS). Uniqueness of the work is that it can be applied to topological features of electron density distributions in QTAIM (Quantum Theory of Atoms In Molecules) and virtual high-throughput screening (VHTS) with largely empirical fields of QTAIM. Moreover, the procedure allowed rapid computations of electronic property in bio-molecules and large molecular datasets. These researchers also developed a RECON algorithm for handling rapid generation of ab initio electron densities and correlation with electronic properties for large molecules [14].

However, more generalization of molecular electrostatic potential in the study of non-covalent interactions was presented by Orozco et al. [15]. They introduced environment effects and non–electrostatic energy considerations in the MEP calculations and provided directions for future research in the field.

Gadre et al. [16] further extended the idea as molecular recognition via electrostatic potential topography by developing various models for weak intermolecular interactions.

Next, in dealing with biopolymers like DNA, Mishra and Kumar [17] made predictions using MEP descriptors and validated profiles by diffraction experiments. Presence of electrophilic and nucleophilic sites in DNA, and accessibility to different atomic and molecular sites to reactants, effect of metal ions and role of dielectrics to solvent media were reliably interpreted using MEP values. MEPs were also shown to provide a better assessment of bond strengths for other molecules with suitable modifications.

In a seminal review, Geerlings et al. [18] compared molecular electrostatic potentials with DFT descriptors for reactivity and combined theoretical concepts and principles with critical evaluations of their applications. Descriptors, such as electrostatic potential, bond density and bond order in molecules and clusters were shown in the applications.

Politzer and Murray [3] explained the significance of MEP in molecular property studies by showing clear relationships between MEP and intrinsic properties. They demonstrated the usefulness of three dimensional visualization of MEP showing how charge distributions of molecules can provide locations of nucleophilic and electrophilic sites along with the size and shape of molecules. The MEP profiles showing both electronic and steric characteristics of molecules soon became invaluable tools for predicting behaviors of complex molecules and found applications in medicinal chemistry.

In a recent study, Orthaber et al. [19] exploited local minima of molecular electrostatic potentials for determining protonation sites in molecules. Since a large number of distinct structures involving hydrogens were found in the mechanism, the researchers developed an automated algorithm for thorough analysis of catalytic processes using MEPs. With similar observations, another research group used MEP-based protonation as a descriptor for automation to get deeper insights on complex organometallic reaction mechanisms. Since conformational freedom and presence of multiple protonation sites were involved in different structures, automation using MEP descriptors for the reaction network was found to be quite useful [20].

In another study, Vetrivel et al. [21] showed how MEP descriptors could be useful in competing haloethynylbenzenes with fluorinated aromatic halogen bond donor functionalities and how the activation energy can affect its relative ranking with established halogen bond donors. They found that MEP values have not only a significant role in the hybridization process of carbon atoms adjacent to aromatic ring halogens in electron-withdrawing carbon atoms but also on polarization of carbon atoms. According to their calculated MEP values, iodopentafluorobenzene was observed to be having double activation resulting in a positive MEP potential (20–40 kJ/mol) at the σ-hole of iodine atom (known to be halogen bond donors) showing clearly the significance MEP descriptors. Furthermore, calculations of MEP values enabled ranking the relative efficiency of halogen bond donors and created opportunities for building strategies for crystal engineering based on hierarchical competition of halogen bonds. Calculated MEPs on individual molecules also led to reliable and practical synthetic routes for cocrystal synthesis and experimental structural designing of solids. Abeysekera et al. [22] and coworkers validated the idea by performing a series of cocrystallization experiments of biimidazole-based symmetric ditopic acceptors. Relative ranking of haloethynyl moieties, fluorinated, and nonactivated halogen bond donors based on calculated molecular electrostatic potentials showed hybridization of carbon atoms adjacent to halogen atoms and thus could overcome the electron-withdrawing effect of fluorine atoms attached an aromatic moiety. This polarization effects of sp-hybridized carbon atoms was reported to play an important role in the synthesis process. These workers [22] further explored a process known as "double activation" by combining the electron-withdrawing capacities of nitro-moieties with the polarizing effect of sp-hybridized carbon atoms for producing highly electrophilic halogen bond donors. From calculated MEPs, the researchers were able to provide a relative ranking of "doubly", "singly," and nonactivated halogen bond donors. Three kinds of halogen bond acceptors: N-based, O-based, and N-/O-based acceptors were found. IR spectra of co-crystallized products were reported to be comparable with spectra of individual reactants.

The intermolecular interactions have a major role in various volumetric properties. These properties influence thermodynamic properties of liquids and their mixtures. The intermolecular interactions involved in liquids and their mixtures are difficult to understand experimentally. However, these interactions can be studied efficiently by theoretical methods, such as DFT calculations. Calculated MEP values can provide an accurate information on the interaction sites. Recently, Kabadi et al. [23] reported a DFT Study of hydrogen bonding effects on molecular volume of water clusters $(H2O)_n$ where $n = 2$–10, Kabadi [24]. Attractive interaction energies of water clusters were found to increase with increase in cluster sizes. Molecular volumes of water clusters were found to increase with increase in the cluster size, and the difference between molecular volume and ideal volume was found to decreases with increase in cluster size confirming the involvement of strong attractive intermolecular interactions in water clusters.

3 Studies from Author's Laboratory on the Application of MEPs

3.1 Anti-inflammatory and Cardiotonic Compounds

The theoretical models presented here are mostly based on density functional and semi-empirical quantum chemical computations on known bioactive compounds or drugs.

One of the earlier studies involved conformational and electrostatic properties of some bipyridine cardiotonics [25]. The next study was on phenyl, -thienyl and -thienyl glycolic acid anti-inflammatory compounds which could provide a correlation between biological activity and electrostatic similarities [26]. Both above studies indicated the importance of rotational barriers in those compounds and influence on electrostatic potential values, consequently their respective biological activities.

In a collaborative study on cardiotonics, focusing specifically on bipyridine cardiotonics: amrinone and milrinone, Kumar et al. [27] extended MEP descriptors to electric field mappings by demonstrating a case study with substituted 3-pyridine-carboxylic acid cardiotonics and establishing their structure–activity relationships [28].

The study on cardiotonics was further extended through higher level of theoretical investigations in collaboration with another research group. In the study, greater conformational flexibility of amrinore over milrinone was observed thus, greater cardiotonic potency of milrinone. Electrostatic potential analyses on the compounds were found to be consistent with observation that planar conformers of both compounds have little influence on cardiotonic activity and reduced conformational flexibility of milrinone was responsible for its greater potent activity [29]. In another work, electronic properties of several novel heterocyclic cardiotonics were computed and MEP values were used to interpret biological activities of the compounds. MEP values of different conformers of the substituted cardiotonics indicated 5-cyano and carbonyl oxygen of the carboethoxy group primarily regulate the cardiotonic potency of the compounds [30]. In continuation of the work, calculated MEPs were used to explain diverse pharmacological properties of γ-aminobutyric acid (GABA) and its mediators. MEP profiles on hydration of GABA indicated a positive potential region on nitrogen atom but hydrophobic region remained unaltered. However, a weak negative potential region was observed around the carbonyl group of lactone ring which appeared to be responsible for binding to GABA receptor sites. Thus, MEP profiles of hydrated GABA and hydrated bicuculline indicated that the later was an antagonist of hydrated GABA [31]. In addition, another study on cardiotonics was reported by the author's group. It was a structural and electrostatic homology study between bypyridine cardiotonics and thyroxine (T_4) hormone. It was investigated using ab initio (3-21G basis set) quantum chemical method. An electrostatic homology between phenolic ring of thyroxine and the substituted milrinone ring was observed. The electrostatic homology and reduced torsional flexibility were implicated for thyromimetic potential towards Ca^{2+}—ATPase [32].

In continuation with the study on these compounds, a specific study was performed with Gadre's group which involved theoretical investigations on cardiotoxicity of anthracycline family of antibiotics. The molecular backbone of these therapeutic agents contains a tetracycline anthraquinone nucleus linked through a glycosidic bond to an amino-sugar. Despite clinical effectiveness, the drugs were known to cause serious cardiotoxicity probably due to inhibition of DNA synthesis. Reactive radicals have been reported to cause damage to DNA or induce lipid peroxidation in the structure of membranes. Theoretical understanding of semiquinones derived from adriamycin was not much known. Since quinonoid moiety of the drug had been reported to play a major role for cytotoxicity and cardiotoxicity, conformational and electrostatic properties of a few simple analogs of adriamycin, such as naphthazarin (5,8-dihydroxy-1,4-naphthoquinone), juglone (5-hydroxy-1,4-naphthoquinone) and naphthoquinone were studied using ab initio RHF-SCF methods with different basis sets for MEP calculations. Two observations that were interesting about the MEP profiles: (a) the depth, extent and relative location around hydroxyl and quinonoid oxygen atoms, (b) a gradual decrease of negative potential over the molecular plane due to the presence and orientation of hydroxyl group in the phenolic part of the molecule. These features are considered to be important for recognition interactions with the receptor. In addition aqueous solvation was also found to have significant influence on the MEP profiles of the compounds [33].

3.2   Mutagenicity

Next, in another collaborative study with Gadre's group, an attempt was made to correlate MEP values of acetaldehyde, nitrous acid and hydroxyl amine with mutagenic and toxicological properties of the compounds using calculations at high levels of quantum theory, such as ab initio SCF using TZ2p, 6-31G* and STO-3G basis sets. Magnitude of negative MEP of all the compounds were found to be correlated with their proton affinities. Since MEPs reflect interaction energy with a positive point charge that is the amount of energy necessary to extract a proton from the proton donor site of the receptor, the correlations were consistent [34].

3.3   Antimalarial Drug Design and Discovery Studies

These studies were performed at the Walter Reed Army Institute of Research (U.S.A.). The goal of the study was to discover potential antimalarial drugs to circumvent the spread of drug-resistant *Plasmodium falciparum* strains in parts of the tropical world. In pursuit of this goal, one of the earliest studies by the author's group was a computational study on 4-aminoquinolinecarbinol amine compounds. Molecular electrostatic potential onto electron density surface of the compounds was computed using ab initio quantum chemical methods. This MEP profiles were found to play a pivotal role for antimalarial activity of the compounds. Large laterally extended negative electrostatic potential by the quinoline nitrogen and absence of negative potential over the molecular plane appeared to play a crucial role for potent antimalarial activity. In addition, the electrostatic features and the MEP values on the electron density surface were also indicated to play an important role in modulating hydrophobicity or lipophilicity of the aminoquinolines and thus, their antimalarial activities. This modeling study was later found to be useful for design and discovery of more efficacious antimalarial aminoquinolines [35].

Next, a theoretical study was performed on structurally diverse synthetic aromatic carbinolamines containing phenanthrene, quinoline, and N-substituted biphenyl rings to find out the functional correlation between their molecular electronic properties, specifically the MEP values and antimalarial potency. Results indicated the aliphatic nitrogen atom and hydroxyl proton in phenanthrene were intrinsically more nucleophilic and electrophilic, respectively than the non-phenanthrene compounds. Hydrogen bonding capacity and overall electrophilic nature of the aromatic ring appeared to play important roles for interaction with the receptor and activity, Bhattacharjee and Karle [36]. Next, theoretical analysis of electronic features of eight cinchona alkaloids and an in vivo metabolite was performed in relation to their in vitro IC_{50} values against both chloroquine sensitive and resistant *P. falciparum* malaria. Results indicated distinguishing features between weakly active epiquinine and epiqinidine which included (a) a higher dipole moment and different direction of electric field, (b) a greater intrinsic nucleophilicity, (c) lower acidity of the hydroxyl proton, (d) a lesser electron affinity of the LUMOs, and (e) a stronger proton affinity of the aliphatic quinuclidine nitrogen atom (determined by MEP calculations) to be responsible for activity of cinchona alkaloids [37].

In continuation with MEP studies on compounds with anti-malarial properties, an interesting study was performed on structural specificity of chloroquine-hematin binding associated with inhibition of hematin polymerization and the growth of malaria parasite. Results of our ab initio 3-21G* level of MEP calculations on several analogues of 4-aminoquinolines and chloroquine were found to be consistent with the experimental observation that favorable enthalpy occurs when π-π interaction takes place between chloroquine and hematin forming the μ-oxo dimer complexes. Alignment of the out-of-plane π- electron density in chloroquine appeared to be responsible for its complexation with hematin to form the hematin μ-oxo dimer at points of intermolecular contact. Electron withdrawing functional groups at the 7-position of the quinoline ring in 4-aminoquinolines were found to be required for activity against both hematin polymerization and growth of parasite. This observation and MEP profiles suggested that chlorine substitution at the position 7 is optimal for the aminoquinolines and appeared to be responsible for antimalarial activity of chloroquine [38]. Another related malarial study was carried out with carbon isosteres of the 4-aminopyridine substructure of chloroquine in order to support the importance of chloroquine hematin binding in the mechanism for antimalarial activity of chloroquine. Three dimensional electrostatic potential profiles beyond van der Waals surfaces and atomic electrostatic charges showed a clear shift of the π- electron density distribution toward the pyridine N atoms relative to chloroquine. This was also reflected from the diminishing atomic electrostatic charge density values in isosteres from chloroquine. Experimental observations compared to chloroquine, hematin binding affinity of the carbon isosteres were found to be much weaker or of no measurable consequence [39].

Another important quantum chemical study was taken up on molecular electronic structures of an important class of antimalarial compounds known as, artemisinin (qinghaosu) and its analogues. Detailed ab initio quantum chemical calculations were performed on artemisinin and eight of its derivatives with complete optimization of geometry of each followed by calculation of their stereoelectronic properties using the 3-21G split valence basis sets. Comparison of results with available in vitro neurotoxicity data could differentiate between analogues with higher and lower neurotoxicity. The results may be summarized as that the least neurotoxic compounds were more polar with an electric field pointing away from the endoperoxide bond, have a higher positive electrostatic potential on the van der Waals surface of the all carbon-containing ring C, a more stable peroxide bond to cleavage, a less negative electrostatic potential by the endoperoxide (Fig. 3a), and a single negative potential region extending beyond the van der Waals surface of the molecule (Fig. 3b). In general, the overall results indicated higher intrinsic lipophilicity (weakly distributed negative electrostatic potentials) is associated with greater neurotoxicity [40].

In relation to peroxides in antimalarial compounds, we performed another study on a series of peroxy ketals. Our results indicated: (a) a shift of the primary focus of recognition interaction from the peroxide bonds, and (2) an increase in the strength of peroxide bond. Thus, source of free radicals in the mechanism of action of the peroxide containing antimalarials would be reduced and likely to be less potent than artemisinin compounds. Therefore, an important tradeoff was found between antimalarial activity and neurotoxicity in peroxide containing antimalarial compounds that could be determined using MEP descriptors [41]. In continuation with artemisinin studies, we focused on another important artemisinin analogue, β- artelinic acid known to have less CNS toxicity in order to clarify the structural requirements of the compounds for less toxicity but retaining antimalarial activity. Using a combination of MEP descriptors, NMR and cyclic voltammetry

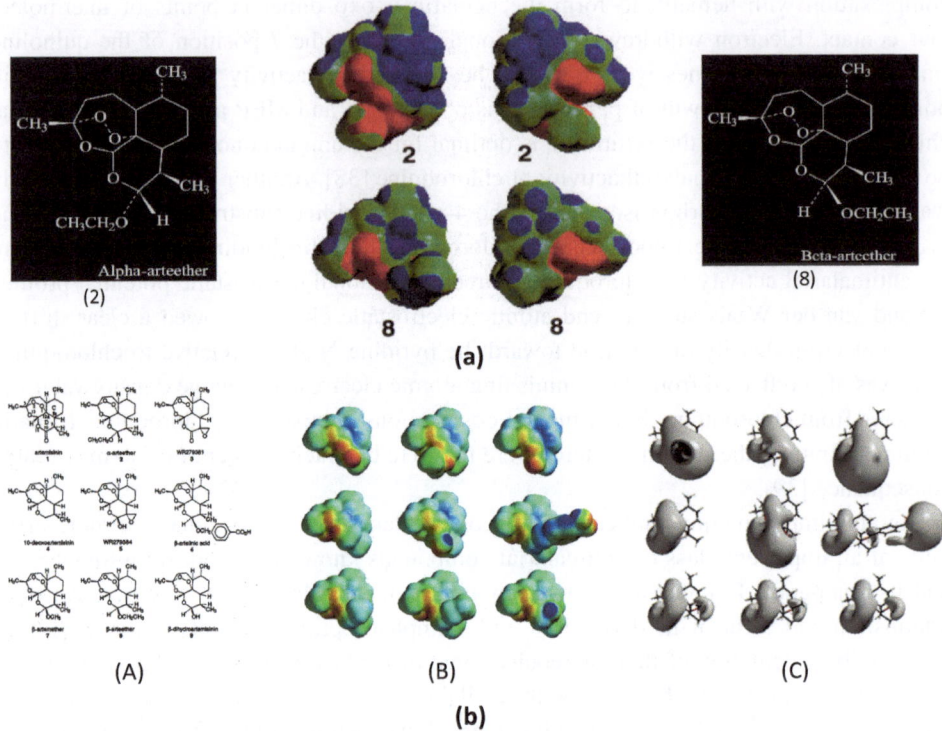

Fig. 3 **a** Discrete bands of MEP profiles. (Left) alfa-arteether, (right) Beta arteether(Blue > 10 kcal/mol; Green = −10 to 10 kcal/mol; Red < −10 kcal/mol). **b** Interaction—pharmacophore model of artemisinin and its analogues for explaining neurotoxicity. (left) structure of artemisinin and its analogues; (middle) MEP profiles on the van der Waals surface of the molecule; (right) MEP profiles beyond van der Waals surface at −5.0 kcal/mol of the molecules

experiments, we observed that the intrinsic nucleophilicity calculated from MEP values of peroxide oxygen atoms remained unchanged and was consistent with results obtained from redox potentials measurements by cyclic voltammetry as well as by NMR experiments. These observations should aid in the design of new artemisinin analogues having potent antimalarial activity and reduced CNS toxicity [42]. In addition, we performed another NMR and MEP analysis on interaction of artelinic acid and artesunic acid with β-cyclodextrin since both these drugs have poor solubility and stability issues in aqueous solution despite showing promise for the treatment of multidrug resistant *Plasmodium falciparum* malaria. This investigation indicated that the electrostatic component of both compounds and β-cyclodextrin plays a dominant role in the complexation process. These results may provide a formulation scheme for artelinic acid, artesunic acid and similar artemisinin derivatives to improve aqueous solubility retaining antimalarial activity [43].

Further structural studies were performed with tetraoxanes to correlate with antimalar-ial activity of compounds where electrostatic potential descriptors were found to play important roles in the mechanism of action. Tetraoxanes studied were methyl-substituted dispiro-1,2,4,5-tetraoxanes and steroidal 1,2,4,5-tetraoxanes [44, 45].

In the next study, we demonstrated the important role of MEP descriptors on chloro-quine antimalarial resistance reversal agents that included imipramine, desipramine, and several of their analogues, some of which fully reversed CQ-resistance, while others were without much effect. By developing electrostatic potential profiles characterized by a localized negative potential region by the side chain nitrogen atom and a large region covering the aromatic ring, we observed a significant role of MEP in the resistance rever-sal of process. The calculated data further revealed that aminoalkyl substitution at the N5-position of the heterocycle and a secondary or tertiary aliphatic aminoalkyl nitrogen atom with a two or three carbon bridges to the heteroaromatic nitrogen (N5) are required for potent "resistance reversal activity". Lowest energy conformers of the compounds were determined and optimized to afford stereoelectronic properties such as molecular orbital energies, electrostatic potentials, atomic charges, proton affinities, octanol–water partition coefficients (log P), and structural parameters. A fairly good correlation was found between resistance reversal activity and intrinsic basicity of the nitrogen atom cal-culated from MEP descriptors at the tricyclic ring system, frontier orbital energies, and lipophilicity. Significantly, most of the structurally diverse CQ-resistance reversal agents were found to be consistent with another 3D QSAR pharmacophore model [46, 47]. Fur-thermore, the possibility for a cation-pi type binding affinity of chloroquine at the receptor site was also assessed using MEP descriptor based interaction with a positive ion [48].

Along with the above studies, another interesting class of antimalarial compounds, 4-azaindolo[2,1-b]quinazoline-6,12-dione (tryptanthrins) were studied [49]. These com-pounds exhibited very strong in vitro activity against Plasmodium falciparum and low cytotoxicity. Due to the structural similarity of the 4-azaindolo[2,1-b]quinazoline-6,12-diones with aminoquinoline moiety of chloroquine and both being known to interact with heme, we performed similar potential interactions between tryptanthrins and heme. A series of six 4-azaindolo[2,1-b] quinazoline-6,12-dione analogs substituted at the 8 or 9- position were synthesized and their hemin binding affinity was determined by 1H NMR methods. Our combined NMR, X-ray crystallography, and MEP studies on the compounds indicated a strong possibility of weak pi-type interactions with hemin. The observation could be a probable mechanism of inhibition of polymerization of hemin lead-ing to death of malaria parasites by the compounds. The molecular electrostatic potential (MEP) profiles of the un-complexed azaindolo [2,1-b]quinazoline-6,12-dione appeared to have guided the site of interaction with the cation. The pi electrons of the D ring in the unsubstituted compound remain unaffected as seen from the large electron distribu-tion of weak electrostatic potentials over this ring and thus the cation in its optimized complexed structure remained over the centroid of the ring. In the methoxy compound, due to the electron donating nature of the substituent, the D ring gets reinforced with electrons, which is clearly noticeable from the large electron distribution. Thus, results of this study indicate that the azaindolo[2,1-b]quinazoline-6,12-dione family of antimalarials may exhibit multiple paths for antimalarial activity, though inhibition of polymerization of hemin seemed crucial for the process. The study should aid the design and discovery of new target specific antimalarials [50].

3.4   Antileishmanial Drug Design and Discovery Studies

In addition to antimalarial studies, our lab at the Walter Reed Army Institute of Research also performed research in the discovery of potential antileishmanial drugs. One of these studies involved theoretical approaches to understand the mechanism of action of a few macrocyclic bisbenzylisoquinoline alkaloids. MEPs calculated on the optimized geometry of these compounds showed that differences in intrinsic nucleophilicity and electrophilicity affect proton donor acceptor ability and thus, influence intermolecular hydrogen bond formation. Large magnitude and precise location of negative and positive electrostatic potentials and ability to form a hole in the three dimensional macrocyclic structure appeared to have a role in potent antileishmanial activity of these compounds [51]. In another mechanistic antileishmanial study, calculated MEP on camptothecin analogues indicated delocalization of positive potential in the methylenedioxy camptothecin analogues which probably contributed towards the affinity of these molecules for DNA. In addition, geometrical and electronic differences between the E ring of camptothecin and its methylenedioxy analogues were noted. One or both of these factors were believed to be responsible for superior biological activity of the methylenedioxy camptothecin analogues [52].

3.5   Insect Repellent Design, Discovery and Mechanism Studies

Along with above MEP descriptor based studies, our lab also contributed the idea of MEP to the understanding of insect repellent mechanism to series of compounds. One of the first such studies was on 31 repellents including DEET (N,N-diethyl-3-methyl benzamide). Interestingly, observed MEP values could be predicted for optimal repellent activity of the compounds. This was reflected in high maximum positive potentials on the surface of the van der Waals surface of the molecules and a negative charge density on the amide nitrogen atom which could be correlated to shorter protection time for repellent activity. Overall, values of electrostatic potential on the van der Waals surface by the amide nitrogen and oxygen atoms, atomic charge at the amide nitrogen atom, and the dipole moment were all found to be in optimal ranges to account for potent repellency [53].

In continuation with insect repellent studies, we performed a molecular similarity analysis between insect juvenile hormone mimics and N,N-diethyl-3-methyl benzamide (DEET). We observed similarity of stereoelectronic attributes of the ester/amide, negative electrostatic potential regions beyond the van der Waals surface and a large distribution weak electrostatic potentials of hydrophobic regions in compounds were responsible for similar interaction with the JH (juvenile hormone) receptor. The electrostatic similarity beyond the van der Waals surface suggested a bioisosterism of the amide group of DEET, its analogues and the JH receptor for molecular recognition to each other [54, 55]. Further studies on the topic was published in the form of a book chapter [56]. In addition another combined streoelectronic study on arthropod repellent activities focusing on MEPs were published in a book chapter [57].

3.6 Reactivators of Organophosphorus (OP)-Inhibited Acetylcholinesterase (AChE) Studies

In another interesting area of research involving discovery of potential non-oxime reactivators of organophosphorus (OP) -inhibited acetylcholinesterase (AChE), we performed a series of studies to show the importance of stereoelectronic properties in the correlation of reactivation efficacy of oximes to OP-inhibited AChE. We found consistency of the properties after developing a statistical pharmacophore model for reactivation of oximes which subsequently enabled us to discover a few potent non-oxime reactivators of OP-inhibited AChE. We showed that location of molecular orbitals and molecular electrostatic potentials were crucial for developing the statistical model.

To begin with, inhibition of cholinsterases with cationic phosphonyl oximes was first studied. It was observed that the stability ranking order of the pyridinium oximes in aqueous medium correlated well with the electronic features mainly with MEP values. This significant study provided a better insight in understanding the major role of cationic pyridinium oxime leaving group in the inhibition reaction [58]. The understanding enabled to go forward for developing a pharmacophore model for virtual screening of compound databases. The goal was discovery of non-oxime reactivators through virtual screening databases, shortlisting of identified non-oximes and experimental testing of them both in vitro and in vivo for efficacy. Indeed, we were successful in identifying several non-oxime reactivators. After in vitro and in vivo evaluations, we shortlisted a few compounds for animal testing to show two of the oximes to have efficacy comparable to 2-PAM used by the U.S. Army to mitigate any OP-inhibited AChE effects in soldiers [59, 60]. In continuation with this area of research, another MEP study was performed against cyclosarin (GF) inhibited AChE to show how pharmacophore was useful for discovery of new class of reactivators [61, 62].

Apart from pharmacophore modeling, other important observations in these studies include specific stereoelectronic profiles, distance between bisquarternary nitrogen atoms of the pyridinium ring in oximes, hydrophilicity, surface area, nucleophilicity of the oxime oxygen, and location of the molecular orbitals on the isosurfaces. All these factors were shown to play an important role in determining efficacy for reactivating GA-inhibited AChE. In summary, affinity for binding to GA-inhibited AChE appeared to require a hydrogen bond acceptor, a hydrogen bond donor at the two terminal regions, and an aromatic ring in the central region of the oximes [63]. Results of this mechanistic study provided a foundation for developing a pharmacophore model of oxime affinity.

Another significant study was reported by the author's group during the time. It was a challenging attempt to find an alternative to atropine that emphasizes M1 (seizure prevention) antagonism but retains minimum M2 (cardiac) and M3 (e.g., eye) antagonism. Since atropine has been in use for seizure prevention quite often, certain issues regarding its use are frequently ignored. Although atropine has a broad spectrum of activity, it is responsible for certain adverse effects for its use, particularly towards the heart and the eye. The goal of the study was to find new compounds which could have the potential as alternatives to atropine. This study indeed allowed us to find out a few potential compounds too using MEP based interaction pharmacophore model [64]. In order to develop an effective alternative to atropine, statistical pharmacophore modeling strategy was adopted that have the characteristics of known M1 subtype-selective compounds. Using the model several antagonists were identified by screening compound databases. MEP profiles provided the foundation of the model showing a few consistencies linked to antimuscarinic activity of newly identified compounds: (a) non-bonded distance between the carbonyl oxygen and the protonated nitrogen atom containing heterocyclic ring, (b) torsion angle between the aromatic ring plane and heterocyclic ring, (c) reactivity indexes (HOMO–LUMO energy difference), (e) specific location MEP potentials, HOMO and LUMO isodensity surfaces, (f) nucleophilic affinity of the nitrogen in the heterocyclic ring, and (g) hydrophobicity of the aromatic moieties. Stereo-electronic requirements for developing the effective statistical pharmacophore model included two hydrogen bond acceptor sites (found by the most nucleophilic sites in MEP profiles near large electron density regions) and aliphatic hydrophobicity found by the large weakly distributed potential regions and an aromatic ring site.

3.7   Antiviral Design and Discovery Studies

For a pursuit of antiviral drug discovery, our group in the Georgetown University collaborated with Howard University, Washington DC, U.S.A. In one of these published studies, we utilized molecular electrostatic potential profiles as interaction pharmacophore model from aminoquinoline antimalarials for in silico search of new antiviral compounds from literature. The search led to identification of several new aminoquinolines and isoquines. Selecting three from these new aminoquinolines, quinacrine (QC), mefloquine (MQ) and N-tert-Butyl isoquine (GSK369796), an experimental evaluation for antiviral efficacy was performed. All the three compounds were found to have potent activity against dengue and zika viruses [65]. For inhibition of DENV2 replication replicon and infectivity (plaque), qRT-PCR assays were used whereas for evaluation of anti-ZIKA activity, plaque assay was used. Our results indicate that all three compounds inhibited dengue and ZIKV infectivity. For dengue, EC50 values of QC 7.09 ± 1.67 MQ 4.36 ± 0.31 GSK369796 3.03 ± 0.35; and for ZIKV, EC50 of 2.27 ± 0.14 mM (QC), 3.95 ± 0.21 mM (MQ) and 2.75 ± 0.09 mM [65, 66].

Ever since novel SARS CoV-2 virus (Covid-19) emerged as a highly infectious human pathogen in the late 2019, discovery of therapeutics against the infection became an urgent need and important objective for antiviral research. Although several successful vaccines became available from early 2021 for prevention from the infection, only two drugs, PAXLOVID™ (nirmatrelvir, ritonavir) from Pfizer and molnupiravir from Merck obtained Emergency Use Authorization for Covid-19 treatments. However, both drugs are reported to have certain shortcomings and issues. Therefore, efforts for discovery of new drugs are continuing particularly, to counter infections from different mutants of Covid-19. In pursuit of the goal, we performed a study [67] focusing on MEP based interaction pharmacophore modeling, specifically on repurposed drugs which were used recently for treatments of Covid-19. Drugs, such as chloroquine, hydroxychloroquine, arbidol, remdesivir, and favipiravir have undergone clinical studies against Covid-19. However, the drugs failed to achieve the highest objective for successful treatment of Covid-19 infection largely because of safety concerns. Our focus was to extract the intrinsic reactivity information of the above drugs by developing accurate MEP profiles using quantum chemical methods and thereby develop a reliable "interaction pharmacophore" model to enable search for effective less toxic anti Covid-19 compounds. From the MEP profiles beyond van der Walls surface of the chloroquine (CQ) and hydroxychloroquine (HCQ) at -20 kJ/mol, an well-defined electrostatic bio-isosteric template was found which was utilized for search of potential new anti-Covid-19 compounds. A literature search using the bio-isosteric template led to the identification of several new aminoquinolines and isoquine analogs. All these compounds showed similarity of electrostatic bio-isosterism with chloroquine and hydroxychloroquine (Fig. 4a and b).

The interaction pharmacophore model showing interaction capacity of isoquines was compared with an inhibitor bound x-ray crystallographic structure of the main protease 3CL-protease (MPro) of Covid-19 to validate the reliability of our model. The interactions were found be consistent in terms for capacities, such as multiple H-bond acceptors (HBA), at least one hydrogen bond donor (HBD), ring aromaticity (RingArom), hydrophobic (Hb) interaction and electrostatic bio-isosterism (similarity for recognition by the target structure) similar to CQ and HCQ. Accordingly, three compounds, (quinacrine, mefloquine and N-tert-Butyl isoquine (GSK369796) were selected for preliminary studies. The results are encouraging and indicate promise for further study with isoquines, particularly with N-tert-Butyl isoquine (GSK369796) as a potential anti-Covid-19 compound. It may be worthwhile to mention here that N-tert-butyl isoquine (GSK369796) is the least

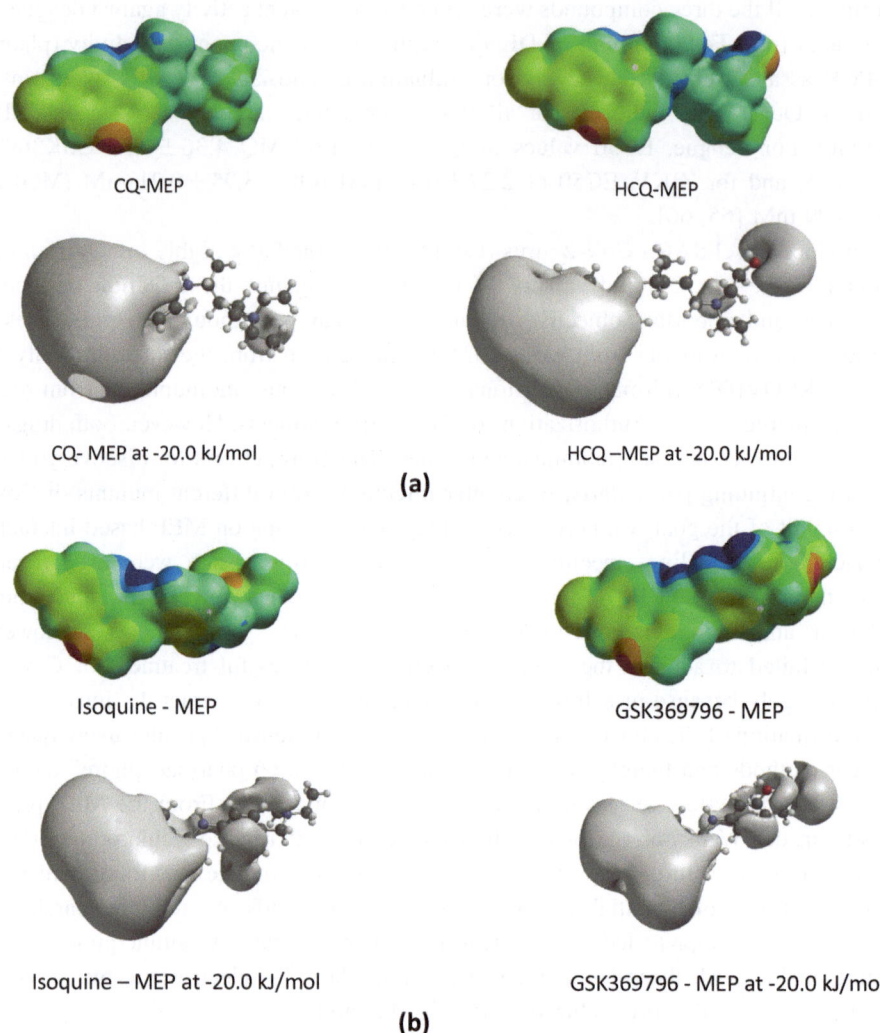

Fig. 4 a MEP profiles onto van der Waals surface (upper) and beyond van der Waals surface (lower). (left) chloroquine (CQ), (right) hydroxy chloroquine (HCQ). **b** MEP profiles onto van der Waals surface (upper) and beyond van der Waals surface (lower). (left) isoquine, (right) N-tert-Butyl isoquine (GSK369796)

toxic amongst the three selected compounds having much higher antiviral therapeutic index compared to the other two. Isoquine analogues including GSK369796 are safer and less toxic than CQ & HCQ because isoquine derivatives are not capable of forming the toxic quinoneimine metabolites via cytochrome due to structural modifications in them by the substitution of a hydroxyl group in the phenyl ring side chain. This Mannich side chain enables them to prevent oxidation into toxic metabolites. It is noteworthy to mention that GSK369796, N-tert-Butyl isoquine, was designed based on chemical, toxicological, pharmacokinetic and pharmacodynamics properties for malaria treatments as part of a public private partnerships between academics at University at Liverpool, UK, Medicine for Malaria Venture (MMV), and GSK Pharmaceuticals. The drug was also investigated in human volunteers (Clinical Trials. gov identifier: NCT00675064 [68, 69]. However, due to evidence of electro-cardiogram changes and lower-than-expected plasma exposure, GSK discontinued GSK369796 after Phase I trial. Nevertheless, isoquine and its derivatives are usually less toxic and thus show promise for further development of analogues to counter Covid-19. In summary, our "interaction pharmacophore" model developed in this study could be useful for further discovery of more potent anti-Covid-19 compounds [70]. A comprehensive list of compounds studied in the author's laboratory, their probable targets, databases used to find them and corresponding references are provided in Table 1.

4   Future Scope

Interaction pharmacophore models and MEP profiles of biologically active molecules (potential drug molecules) could be useful for augmenting machine learning and AI (artificial intelligence) tools perhaps as "finger prints" for search of large commercially available compound databases containing millions of compounds. The procedure would likely to further improve the efficacy of large scale virtual screening to leverage the HTS (High throughput screening) for rapid drug discovery. However, development of computer algorithms and reliable software are required for the purpose which are needed to be worked out. Although the procedure may require some initial investment but eventually results from the development could be both cost and time effective for the medicinal and drug discovery chemists in pharmaceutical industries.

5   Concluding Remarks

Ever since MEP could be reasonably well calculated using density functional theory as a simplified analytical expression, the unique molecular property started to evolve further with applications in multiple directions, such as for describing chemical reactions and applications to topological features of electron density distributions in QTAIM (Quantum Theory of Atoms In Molecules) and to virtual high-throughput screening (VHTS) with largely empirical fields of QTAIM. Since molecular electrostatic potentials (MEPs), their three dimensional profiles and descriptors provide useful representation of intrinsic reactivity of molecules, soon they found application in drug discovery programs and medicinal chemistry. MEP profiles not only enable visualization and charge distributions and other charge related properties but also provide an invaluable tool for predicting the

Table 1 Shows compounds studied, their probable targets, databases used and references

Drugs and potential active compounds identified by database searches	Targets	Databases searched	References
1. Bipyridine cardiotonics	Act as inhibitors of phosphodiesterase fraction II; Ca^{2+}—ATPase, and DNA	Mechanistic study from literature data	Bhattacharjee [25, 26], Bhattacharjee et al. [29, 33], Guha et al. [31], Kumar et al. [27, 28], Majumdar et al. [30]
2. Mutagenicity	DNA inter-strand crosslinks, but not DNA–protein, mainly cellular	Aldehydes	Bhattacharjee et al. [34]
3. Antimalarial aminoquinolines	Erythrocytic stage of malaria infection	Aminoquinolines phenanthrenes and non-phenanthrenes	Bhattacharjee and Karle [35, 36]; Karle and Bhattacharjee [37]
4. Chloroquine	Hematin	Mechanistic studies on structural specificity	Vippagunta et al. [38], Cheruku et al. [39]
5. Artemisinin and its derivatives	Plasmodium falciparum Heme Detoxification pathway/neurotoxicity	Artemisinin, peroxy ketals and tetraoxanes database	Bhattacharjee and Karle [40, 41], Bhattacharjee et al. [42], Hartell et al. [43], McCullough et al. [44], Bhattacharjee et al. [45]
6. Chloroquine antimalarial resistance reversal agents	Cellular targets	Imipramine, desipramine etc. database	Bhattacharjee et al. [46, 47], Bhattacharjee [48]
7. 4-azaindolo[2,1-b]quinazoline-6,12-diones (Tryptanthrins)	Unknown target	Walter reed army chemical information database (WR-CIS)	Bhattacharjee et al. [49], Hicks et al. [50].

(continued)

Table 1 (continued)

Drugs and potential active compounds identified by database searches	Targets	Databases searched	References
8. Macrocyclic bisbenzylisoquinoline alkaloids and camptothecin analogs	Topoisomerase I (Top1)	Mechanistic studies on macrocyclic alkaloids and camptothecin and its methylenedioxy analogues	Bhattacharjee et al. [51], Werbovetz et al. [52].
9. DEET (N,N-diethyl-3-methyl benzamides)	May be inhibition of cholinesterases in insect and mammalian nervous systems by the insect repellent DEET	Insect repellent database	Ma et al. [53].
10. DEET analogues	JH (juvenile hormone) receptor	Insect repellent database	Bhattacharjee et al. [54, 55], Bhattacharjee et al. [56], Gupta and Bhattacharjee [57].
11. Non-oxime reactivators	Organophosphorus (OP)-inhibited acetylcholinesterase (AChE)	WR-CIS	Ashani et al. [58], Bhattacharjee et al. [59, 60], . Bhattacharjee et al. [61, 62].
12. GA-inhibited AChE oxime reactivators	Mechanism for GA-inhibited AChE reactivation	Reactivators	Bhattacharjee et al. [63].
13. Alternatives to atropine M1 subtype-selective compounds	M1 (seizure prevention) antagonism but retains minimum M2 (cardiac) and M3 (e.g., eye) antagonism	WR-CIS	Bhattacharjee et al. [64].

behavior of complex molecules. MEP and its profiles are particularly useful for medicinal and organic chemists for synthesis and prediction of reaction paths as well as prediction of properties for synthesized molecules. The statistical pharmacophore built with the foundation of MEP became a powerful tool for rapid search of compounds. Since MEP is an experimentally determinable property containing a wealth of reliable information on intrinsic molecular and material interaction capabilities, it is commonly known as "interaction pharmacophore". The present chapter first introduced the theoretical foundation of MEP and its mathematical expressions followed by various earlier application studies in material sciences, solvents and their complexities, and environmental effects. Finally, the chapter described MEP applications in drug design and discovery studies ranging from mutagenicity and carcinogenicity to studies on cardiotonic, anti-inflammatory, anti-malarial, anti-leishmanial, insect repellents, organo-phosphorous poisoning and antiviral compounds including identification of potential anti Covid-19 compounds from author's laboratory [70]. In summary, interaction pharmacophore modeling from MEP profiles demonstrates a rational approach for discovery of potential therapeutics based on intrinsic reactivity and structure–activity relationships of known bioactive compounds.

References

1. Scrocco E, Tomasi J (eds) (1973) Topics in current chemistry. Springer, Berlin
2. Murray JS, Sen K (eds) (1996) Molecular electrostatic potentials—concepts and applications. Elsevier Science, Hardcover ISBN: 9780444823533, eBook ISBN: 9780080536859
3. Politzer P, Murray JS (1991) Review in computational chemistry. In: Lipkowitz KB, Boyd DB (eds) VCH Publishers, New York (Chapter 7)
4. Naray-Szabo G, Ferenzy GG (1995) Molecular electrostatics. Chem Rev 95:829–847
5. Karelson M, Lobanov VS, Katritzky AR (1996) Quantum-chemical descriptors in QSAR/QSPR studies. Chem Rev 96:1027–1044, and references cited therein
6. Scrocco E, Tomasi J (1978) Advances in quantum chemistry. In: Lowdin P (ed) vol II in, Academic Press, New York, 1978, p 115
7. Naray-Szabo G (1993) Analysis of molecular recognition: steric, electrostatic and hydrophobic complementarity. J Mol Recognit 6:205–210
8. Leach AR, Gillet VJ, Lewis RA, Taylor R (2010) Three-dimensional pharmacophore methods in drug discovery. J Med Chem 539–558
9. Güner OF (ed) (2000) Pharmacophore, perception, development, and use in drug design. University International Line (IUL Biotechnology Series, San Diego
10. Lee PJ, Bhonsle JB, Gaona HW, Huddler DP, Heady TN, Kreishman-Deitrick M, Bhattacharjee A, McCalmont WF, Gerena L, Lopez-Sanchez M, Roncal NE, Hudson TH, Johnson JD, Prigge ST, Waters NC (2009) Targeting the fatty acid biosynthesis enzyme, β-Ketoacyl–Acyl carrier protein synthase III (PfKASIII), in the identification of novel antimalarial agents. J Med Chem 52:952–963
11. Tomasi J, Cammi R, Mennucci B, Cappellia C, Corni S (2002) Phys Chem Chem Phys 5697–5712
12. Leboeuf M, Köster AM, Jug K, Salahub DR (1999) Topological analysis of the molecular electrostatic potential. J Chem Phys 111: 4893. https://doi.org/10.1063/1.479749

13. Breneman CM, Martinov M (1996) The use of electrostatic potential fields in QSAR and QSPR. In: Murray JS, Sen K (eds) Molecular electrostatic potential: concept and applications. Elsevier, Amsterdam, pp 143–179
14. QTAIM in Drug Discovery and Protein Modeling. Available from https://www.researchgate.net/publication/285395791. QTAIM in Drug Discovery and Protein Modeling. Accessed 06 Jul 2022
15. Orozco M, Luque FJ (2000) Theoretical methods for description of the solvent effect in biomolecular systems. Chem Rev 100:4187–4225
16. Gadre SR, Bhadane PK, Pundlik SS, Pingale SS, Subhash S (1996) Molecular recognition via electrostatic potential topography. Theoret Computat Chem 3: 219–255. ISSN 1380-7323. http://linkinghub.elsevier.com/retrieve/pii/S13807
17. Mishra PC, Kumar A (1996) Molecular electrostatic potentials and fields: hydrogen bonding, recognition, reactivity and modelling. Theo & Comput Chem (Elsevier), 257–296
18. Geerlings P, De Proft F, Langenaeker W (2003) Conceptual density functional theory. Chem Rev 103:1793–1873
19. Orthaber A, Karnahl M, Tschierlei S, Streich D, Stein M, Ott S (2014) Dalton Trans 43:4537–4549
20. Grimmel SA, Reiher M (2019) The electrostatic potential as a descriptor for the protonation propensity in automated exploration of reaction mechanisms. Faraday Discuss 220:443–463
21. Vetrivel R, Takaba H, Katagiri M, Kubo M, Miyamoto A (1996) Computational studies on the design of synthetic sorbents for selective adsorption of molecules. In: Studies in surface science and catalysis (eds) Elsevier, Amsterdam, Chapter 1.1., 99, pp 3–30
22. Abeysekera AM, Averkiev BB, Sinha AS, Aaker"oy CB (2021) Establishing halogen-bond preferences in molecules with multiple acceptor sites. ChemPlusChem, https://doi.org/10.1002/cplu.202100102
23. Kabadi EM, Pingale SS (2019) Int J Res Anal Rev 6:186
24. Kabadi EM (2021), Ph.D. Thesis Title: Theoretical study of organic molecular clusters: structures, properties and reactivity -2021, department of chemistry savitribai phule Pune University Pune-411 007, India
25. Bhattacharjee AK (1990) Theoretical conformational study of the molecular structures of some bipyridine cardiotonics. Proc Indian Acad Sc (Chem Sci) 102:159–162
26. Bhattacharjee AK (1991) Molecular electronic structure of derivatives of some model anticholinergic and anti-inflammatory compounds: a theoretical conformational and electrostatic potential study. Ind J Chem 30B:991–998
27. Kumar A, Bhattacharjee AK, Mishra PC (1991) Electric field mapping of bipyridine cardiotonics: amrinone and milrinone. J Mol Structure (Theochem) 251:359–366
28. Kumar A, Bhattacharjee AK, Mishra PC (1992) Electric field mapping of some substituted 3-pyridine-carboxylic acid cardiotonics and the possible structure-activity relationships. Int J Quantum Chem 43:579–589
29. Bhattacharjee AK, Majumdar D, Guha S (1992) Theoretical studies on the conformational properties and pharmacophoric pattern of several bipyridine cardiotonics. J Chem Soc Perkin Trans 2:805–809
30. Majumdar D, Bhattacharjee AK, Das KK, Guha S (1993) Conformational and electronic properties of several novel heterocyclic cardiotonics: a theoretical approach. J Mol Structure (Theochem) 288:41–53
31. Guha S, Majumdar D, Bhattacharjee AK (1992) Molecular electrostatic potential: a tool for the prediction of pharmacophoric pattern of drug molecules. J Mol Structure (Theochem) 256:61–74

32. Bhattacharjee AK (1997) Structural and electrostatic homology between some bipyridine cardiotonics and thyroxine (T4) hormone: an ab initio theoretical study. Indian J Chem Sec B 36B:938–942

33. Bhattacharjee AK, Pundlik SS, Gadre SR (1997) Conformational and electrostatic properties of naphthazarin, juglone and naphthoquinone: an ab initio theoretical study. Cancer Invest 15(6):531–541

34. Bhattacharjee AK, Pundlik SS, Gadre SR (1995) An ab initio topographical investigation on the molecular electrostatic potential of some chemical mutagens. Curr Sci 69:58–62

35. Bhattacharjee AK, Karle JM (1996) Molecular electronic properties of a series of 4- quinolinecarbinolamines define antimalarial activity profile. J Med Chem 39:4622–4629

36. Bhattacharjee AK, Karle JM (1998) Functional correlation of molecular electronic properties with potency of synthetic carbinol antimalarial agents. Bioorg & Med Chem 6:1927–1933

37. Karle JM, Bhattacharjee AK (1999) Stereoelectronic features of the cinchona alkaloids determine their differential antimalarial activity. Bioorg & Med Chem 7:1–6

38. Vippagunta SR, Dorn A, Matile H, Bhattacharjee AK, Karle JM, Ellis WY, Ridley RG, Vennerstrom JL (1999) Structural specificity of chloroquine-hematin binding related to inhibition of hematin polymerization and parasite growth. J Med Chem 42:4630–4639

39. Cheruku SR, Maite S, Dorn A, Scorneaux B, Bhattacharjee AK, Ellis WY, Vennerstrom JL (2003) Carbon isosteres of the 4-aminopyridine substructure of chloroquine: effects on pKa, hematin binding, inhibition of hematin polymerization and parasite growth. J Med Chem 46:3166–3169

40. Bhattacharjee AK, Karle JM (1999) Stereoelectronic properties of antimalarial artemisinin analogues in relation to neurotoxicity. Chem Res Toxicol 12:422–428

41. Bhattacharjee AK, Karle JM (1999) Role of molecular electronic properties of some novel antimalarial cyclic peroxy ketals in relation to potency and potential toxicity. Mol Eng 8:391–402

42. Bhattacharjee AK, Skanchy DJ, Hicks RP, Carvalho KA, Chmurny GN, Klose JR, Scovill JP (2004) Structure of β-artelinic acid clarified using NMR analysis, molecular modelling & cyclic voltammetry, and comparison with α-artelinic acid and β-Arteether. Internet Elec J Mol Design 3:55–72

43. Hartell MG, Hicks RP, Bhattacharjee AK, Kozer B, Carvolho K, van Hamont JE (2004) NMR and molecular modeling analysis of the interaction of the anti- malarial drugs artelinic acid and artesunic acid with beta-cyclodextrin. J Pharm Sci 93:2076–2089

44. McCullough KJ, Wood JK, Bhattacharjee AK, Dong Y, Kyle DE, Milhous WK, Vennerstrom JL (2000) Methyl-substituted dispiro-1,2,4,5-tetraoxanes: correlations of structural studies with antimalarial activity. J Med Chem 43:1246–1249

45. Bhattacharjee AK, Carvalho KA, Opsenica D, Šolaja BA (2005) Structure-activity relationship study of steroidal 1,2,4,5-tetraoxane antimalarials using computational procedures. J Serb Chem Soc 70:329–345

46. Bhattacharjee AK, Kyle DE, Vennerstrom JL, Milhous WK (2002) A 3D QSAR pharmacophore model and quantum chemical structure activity analysis of chloroquine(CQ)-resistance reversal. J Chem Info Comput Sci 42:1212–1220

47. Bhattacharjee AK, Kyle D, Vennerstrom JL (2001) Structural analysis of chloroquine-resistance reversal by imipramine analogs. Antimicrob Agents Chemother 45:2655–2657

48. Bhattacharjee AK (2001) Assessment of cation-pi binding affinity of the aromatic ring in several chloroquine analogs using Ab initio quantum chemical (6–31G**) method. J Mol Struct (Theochem) 549:27–37

49. Bhattacharjee AK, Hartell MG, Nichols DA, Hicks RP, Stanton B, Van Hamont JE, Milhous WK (2004) Structure-activity relationship study of antimalarial indolo [2,1-b]quinazoline-6,12-diones (tryptanthrins). Three dimensional pharmacophore modeling and identification of new antimalarial candidates. European J Med Chem 39:59–67

50. Hicks RP, Nichols DA, DiTusa CA, Sullivan DJ, Hartell MG, Koser BW, Bhattacharjee AK (2005) Evaluation of 4-Azaindolo[2,1-b]quinazoline-6,12-diones' interaction with hemin and hemozoin: a spectroscopic, x-ray crystallographic and molecular modeling study. Internet Elec J Mol Design 4:751–764

51. Bhattacharjee AK (1999) In vitro antileishmanial activity of some natural bisbenzylisoquinoline alkaloids could be correlated with their calculated molecular electronic properties. Int J Quant Chem 75:995–1002

52. Werbovetz KA, Bhattacharjee AK, Brendle JJ, Scovill JP (2000) Stereoelectronic features modulating the biological activity of camptothecin analogs. Bioorg & Med Chem 8:1741–1747

53. Ma D, Bhattacharjee AK, Gupta RK, Karle JM (1999) Predicting mosquito repellent potency of DEET analogs from molecular electronic properties. Am J Trop Med Hyg 60:1–6

54. Bhattacharjee AK, Ma D, Karle JM, Gupta RK, Molecular similarity analysis between insect juvenile hormone mimics and N,N-diethyl-3-methyl benzamide (DEET) analogs may aid the design of novel insect repellents. J Mol Recognit 13:213–220

55. Bhattacharjee AK, Gupta RK (2005) Analysis of molecular stereoelectronic similarity between N, N-diethyl-m-toluamide (DEET) analogs and insect juvenile hormone to develop a model pharmacophores for insect repellent activity. J Am Mosquito Control Assoc 21:23–29

56. Bhattacharjee AK (2013) In silico stereoelectronic profile and pharmacophore similarity analysis of juvenile hormone, juvenile hormone mimics (IGRs) and insect repellents may aid discovery and design of novel arthropod repellents. In: Devillers J (ed) Juvenile Hormones and juvenoids: modeling biological effects and environmental fate. Series: QSAR in Environmental and Health Sciences, Taylor & Francis, CRC Press, 13:297–331

57. Gupta RK, Bhattacharjee AK (2006) Discovery and design of new arthropod/insect repellents by computer-aided molecular modeling. In: Debbon M, Frances SP, Strickman D (eds) Insect repellents: principles, methods, & use. Taylor & Francis, CRC Press, pp 195–228

58. Ashani Y, Bhattacharjee AK, Leader H, Saxena A, Hinrichs C, Doctor BP (2003) Inhibition of cholinesterases with charged organophosphates reveal special properties of cationic aromatic leaving groups. Biochem Pharmacol 66:191–202

59. Bhattacharjee AK, Marek E, Le HT, Gordon RK (2012) Discovery of non-oxime reactivators using an in silico pharmacophore model of oxime reactivators of OP-inhibited acetylcholinesterase. Euro J Med Chem 49:229–238

60. Bhattacharjee AK, Marek E, Le HT, Ratcliffe R, DeMar JC, Pervitsky D, Gordon RK (2015) Discovery of non-oxime reactivators using an in silico pharmacophore model of reactivators for DFP-inhibited acetylcholinesterase. Euro J Med Chem 90:209–220

61. Bhattacharjee AK, Musilek K, Kuča K (2013) In silico pharmacophore modeling on known pyridinium oxime reactivators of cyclosarin(GF) inhibited AChE to aid discovery of potential more efficacious novel non-oxime reactivators. Curr Comput Aided Drug Design (CCADD) 9:402–411

62. Bhattacharjee AK, Musilek K, Kuča K, Gordon RK (2012) An in silico stereo-electronic comparison of conventional pyridinium oximes and k-oximes for organophosphate (OP) poisoning. Med Chem 8:1–15

63. Bhattacharjee AK, Kuča K, Musilek K, Gordon RK (2010) In silico pharmacophore model for tabun-inhibited acetylcholinesterase (AChE) reactivators: a study of their stereoelectronic properties. Chem Res Toxicol 23:26–36

64. Bhattacharjee AK, Pomponio J, Evans SA, Pervitsky D, Gordon RK (2013) Discovery of sub-type selective muscarinic receptor antagonists as alternatives to atropine using in silico pharmacophore modeling and virtual screening methods. Bioorg & Med Chem 21:2651–2662

65. Balasubramanian A, Teramoto T, Kulkarni A, Bhattacharjee AK, Padmanabhan R (2017) Antiviral activities of selected antimalarials against dengue virus type and Zika virus. Antiviral Res 137:141–150

66. Bhattacharjee AK (2019) Discovery of anti-zika drugs using in silico pharmacophore modeling in Zika virus. In: Basak SC, Bhattacharjee AK, Nandy A (eds) Zika virus surveillance, vaccinology, and ant-zika drug discovery—computer assisted strategies to combat the menace, Nova Medicine & Health. Nova Science Publishers Inc., NY, pp 39–73

67. Bhattacharjee AK (2021) Discovery of potential anti-COVID-19 (SARS-CoV-2) compounds using interaction-pharmacophore modeling. J Med Chem Drug Design 1–102

68. O'Neill PM, Barton VE, Ward SA, Chadwick J (2012) 4-aminoquinolines: chlo-roquine, amodiaquine and next-generation analogues. In: Staines HM, Krishna S (eds) Treatment and prevention of malaria. Springer Basel AG. https://doi.org/10.1007/978-3-0346-0480-2_2

69. O'Neill PM, Mukhtar A, Stocks PA, Randle LE, Hindley S, Ward SA, Storr RC, Bickley JF, O'Neil IA, Maggs JL, Hughes RH, Winstanley PA, Bray PG, Park BK (2003) Isoquine and related amodiaquine analogues: a new generation of improved 4-aminoquinoline antimalarials. J Med Chem 46:4933–4945

70. Bhattacharjee AK (2024) A review on recent theoretical approaches made in the discovery of potential Covid-19 therapeutics. J Math Chem. https://doi.org/10.1007/s10910-024-01626-4

Network-Based Molecular Descriptors for Protein Dynamics and Allosteric Regulation

Ziyun Zhou, Lorenza Pacini, Laurent Vuillon, Claire Lesieur, and Guang Hu

Abstract

Similar to mathematical chemistry, topology is at the very heart of systems biology. The rapidly increasing availability of protein structure data and network-based technologies allow insights into the biological function of proteins from topology. In this chapter, we summarized some network-based descriptors to study protein dynamics and allosteric regulation, including structure-based network descriptors and dynamic descriptors based on elastic network models. The applications of these network-based molecular descriptors in predicting functional sites and investigating allosteric regulation are illustrated by two case studies: DNMT1 and SARS-CoV-2 spike protein. We argued that network-based molecular descriptors prove to be a toolbox to study protein structures and dynamics, bridging the fields of mathematical chemistry and systems biology.

Z. Zhou · G. Hu (✉)
Center for Systems Biology, Department of Bioinformatics, School of Biology and Basic Medical Sciences, Soochow University, Suzhou 215123, China
e-mail: huguang@suda.edu.cn

L. Pacini · C. Lesieur (✉)
UMR5005, University Lyon, CNRS, INSA Lyon, Ecole Centrale de Lyon, Université Claude Bernard Lyon 1, Villeurbanne, France
e-mail: claire.lesieur@ens-lyon.fr

Institut Rhônalpin des Systèmes Complexes, IXXI-ENS-Lyon, Lyon, France

L. Vuillon
LAMA, University of Savoie Mont Blanc, CNRS, LAMA, Le Bourget du Lac, France

© The Author(s), under exclusive license to Springer Nature Switzerland AG 2025
S. C. Basak (ed.), *Mathematical Descriptors of Molecules and Biomolecules*, Synthesis Lectures on Mathematics & Statistics, https://doi.org/10.1007/978-3-031-67841-7_8

Keywords

Protein structure networks • Elastic network modes • Molecular descriptors •
Functional sites • DNMT • SARS-CoV-2

1   Introduction

Recent progress in protein structure prediction by AlphaFold2 [25] and RoseTTAFold [4]
has opened new avenues to decipher biological functions from the perspective of structural biology based on the proteomics level. To date, the biological function annotation
of proteins is largely based on sparse experimental technology, while available computational tools are usually based on the analysis of sequence conservation [44]. The emerging
paradigm of functional proteomics by the quantification of protein dynamics and the prediction of functional sites is coming to age in the post-Alphafold era [5, 55]. However,
the barriers to entry have now been lowered.

Understanding the mechanisms of protein function is indispensable for many biological
applications, such as protein engineering and drug design. Allosteric behavior is central to
the function of many proteins, such as the regulation of gene transcription and the activities of enzymes and cell signaling [16]. Mutations, posttranslational modification (PTM),
and ligand binding are three main operations to supervise allostery and hence regulate
function. As an intrinsic but elusive property, allostery is a ubiquitous phenomenon where
binding or disturbing a distal site in a protein can functionally control its protein activity
and is considered as the "second secret of life" [36]. The unifying theme and overarching
goal of allosteric regulation studies in recent years have been integration between emerging experiments and computational approaches and technologies to advance quantitative
characterization of allosteric mechanisms in proteins [56]. Despite significant advances,
challenges still exist, including the quantitative characterization of allosteric states and the
prediction of functional sites [49].

Recent years have shown that protein dynamics and allosteric regulation are defined
by their topological architectures [29, 58]. As such, emerging network tools provide new
insights into understanding protein dynamics and allosteric effects. Considering a protein
molecule as a network or a graph, the structure network-based molecular descriptors can
not only predict critical nodes as functional residues [53], but also capture network signaling efficiency and explain allostery in terms of signal transmission [11]. Allostery is
well-known as a dynamic-driven process, and thus, the molecular descriptors describing
protein dynamics and allosteric signal transmission need to be established. The directly
solved way is to construct network models relying on molecular dynamics simulations
rather than solely dependent on static protein structure [41]. On the other hand, molecular
descriptors have been developed based on elastic network models (ENMs) that provide
efficient methods for investigating the intrinsic dynamics and allosteric communication
pathways in proteins [24].

Over the last few years, network-based molecular descriptors have significantly advanced with respect to their robustness, performance, and scope [32], connecting mathematical chemistry and systems biology [2]. With time, both structure and dynamic descriptors have been developed, which measure functionally important properties of proteins such as allosteric regulation. In this chapter, we will survey the network-based molecular descriptors regarding protein dynamics and allosteric regulation, along with the bioinformatics tools developed. Then, the promising avenues of these molecular descriptors will be represented in the case of the prediction of disease-related mutations, PTMs, and ligand-binding sites, as well as the quantification of allosteric regulation.

2 Molecular Descriptors Based on Structural Networks

In network theory, the protein 3D structure is represented as a graph, with individual residues/amino acids as nodes, and edges defined based on a distant cut-off (Euclidian distances) or other non-covalent interactions [54]. Several types of structure-based networks have been proposed, and we present four of them here: the protein contact network (PCN), the protein structure network (PSN), the amino acid contact energy network (AACEN) and the amino acid network (AAN). We also provide the different molecular descriptors obtained from classical network measures, listed in Fig. 1, that are used to analyze protein structures in terms of quantifying allosteric regulation and predicting functional sites.

The mathematical representation of the PCN is the adjacency matrix [10], defined as:

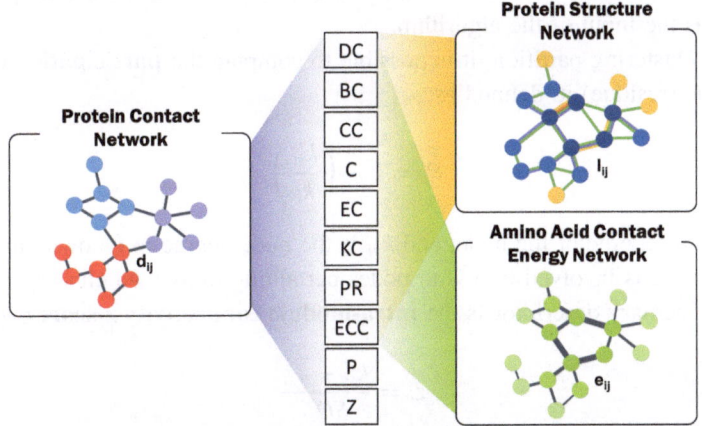

Fig. 1 Structure network-based descriptors. DC: Degree centrality, BC: Betweenness centrality, CC: Closeness centrality, C: Clustering centrality, EC: Eigenvector centrality, KC: Katz centrality, PR: PageRank, ECC: Eccentricity, P_i: participation coefficient, Z_i: intra-module connectivity z-score

$$A_{i,j} = \begin{cases} A_{i,j} = 1 & \text{if } d_{ij} \in \text{cutoff} \\ A_{i,j} = 0 & \text{otherwise} \end{cases} \tag{1}$$

The basic descriptor of the network connectivity is the **node degree k_i** defined as:

$$k_i = \sum_j Ad_{ij} \tag{2}$$

The functional properties of proteins rely on their modularity, that is, different modules (domains) inside the protein structure interact with each other and with the surrounding environment to explicate its function. This interaction between regions of the protein structure is at the very basis of the allosteric mechanism, which in turn is responsible for the regulatory nature of protein activity.

The network formalism also allows the identification of functional modules inside the protein molecule linked to the allosteric pathways throughout the molecule. The functional modules can be obtained by applying the network clustering based on the spectral decomposition of the Laplacian matrix, defined as:

$$L = D - A \tag{3}$$

where D is the degree matrix, i.e. a diagonal matrix whose generic non null element $D_{ii} = k_i$. Applying eigenvalue decomposition to the Laplacian L, the network clustering is a binary, hierarchical method based on the Fiedler vector v_2, i.e., the eigenvector corresponding to the second minor eigenvalue. The sign of its components assigns residues to two different clusters. The final result is a so-called "crispy" clustering partition, meaning a partition of nodes (residues) to different clusters. The number of clusters (power of two) is assigned as the input of the algorithm.

After the clustering partition, it is possible to compute the **participation coefficient P** for each node (residue) it, defined as:

$$P_i = 1 - \left(\frac{k_{si}}{k_i}\right)^2 \tag{4}$$

k_i is the overall degree of the node, and k_{si} is the node degree in its own cluster (number of links the node is involved into with nodes pertaining to its own cluster).

A complementary descriptor is the **intra-module connectivity z-score z**, defined as:

$$z_i = \frac{k_{si} - \overline{k}_{si}}{SD_{si}} \tag{5}$$

where $\overline{k}$ and SD are the average value and the standard deviation of degree k extended to the whole network, respectively. The descriptor z catches the attitude of nodes to preferentially connect with nodes in their own clusters; z strongly correlates with node degree, so high z residues are mostly responsible for global protein stability. Except for MD-TASK as the well-used Python programming tool [8], the most recently developed bioinformatics tool for the analysis of molecular descriptors based on PCN is PCN-Miner [21].

The second type of structure-based network, called PSN here, is an interaction-based network mode for protein structures [23]. The theory of such a PSN is rooted in statistical physics. The importance of edges in PSNs is quantified by the interaction strength I_{ij}, which is the side-chain interactions based on the number of atom–atom contacts (n_{ij}), defined as

$$I_{ij} = \frac{n_{i,j}}{\sqrt{N_j N_i}} \times 100 \tag{6}$$

where N_i and N_j are normalization values of residues i and j. AACENs is the third type of structure network, in which edges are defined by the environment-dependent residue contact energy between two nodes.

$$e_{ij} = -\ln(N_{ij} N_{00} C_{i0} C_{j0} / N_{i0} N_{j0} C_{ij} C_{00}) \tag{7}$$

where N_{ij}, N_{i0}, N_{j0}, and N_{00} are the contact numbers from the known structures, and C_{ij}, C_{i0}, C_{j0}, and C_{00} are the corresponding parameters expected in a reference state.

The standalone and web-based tool called PSN tools was recently developed for the calculation and post-processing of PSN analyses based on either on single structures or on conformational ensembles [17]. An R package [51] and a web server [52] were developed for the construction and analysis of AACENs.

Based on PSNs and AACENs, except for the K_i, some common network centralities have been defined [13]. The **betweenness centrality (BC)** was defined as the number of times residue i was included in the shortest path between each pair of residues in the protein, normalized by the total number of pairs. From the protein point of view, BC ranks amino acid nodes according to the nodes involvement in all the paths between any two amino acid positions in the structure. Hence an amino acid node with high BC can be viewed as a crossroad, a topology which might have a key role for the protein. It is calculated by

$$BC_i = \sum_{j,k \in N, j \neq k} \frac{n_{jk}(i)}{n_{jk}} \tag{8}$$

where n_{jk} is the number of shortest paths connecting j and k, while $n_{jk}(i)$ is the number of shortest paths connecting j and k and passing through i. The **closeness centrality (CC)** for a node was calculated by the reciprocal of the average shortest path length, which can be calculated as follows:

$$CC_i = \frac{(n-1)}{\sum_{k \in N, \ k \neq m} L(m, k)} \tag{9}$$

where N is the set of all modes and n is the number of nodes in the network. The **clustering coefficient (C)** measures the degree to which nodes tend to cluster together and is defined as:

$$C_i = \frac{2e_i}{K_i(K_i - 1)} \tag{10}$$

where K_i is the degree of node i and e_i is the number of connected pairs between all neighbors of i. C_i measures the number of realized triangles between a node i and its neighbors compared to the theoretical number of possible triangles involving i and its neighbors; hence it measures the importance of triangular interactions for a given node, which may impact the mechanisms by which the position can cope with mutation for example.

In addition to these classical network measures that are used as molecular descriptors, some newly developed centrality measures have been introduced for analyzing protein dynamics and allosteric regulation [46], which can be calculated by MDM-TASK-web [47]. **Eigenvector centrality (EC)** is an extension of degree centrality. It assigns node importance by solving for the dominant unit eigenvector EC of the adjacency matrix A. The eigenvector decomposition method that can be used to determine EC, as the following:

$$EC_i = \epsilon^{-1} \sum_{j=1}^{n} A_{ij} c_j \tag{11}$$

The EC is a measure of how well connected a node is to other well-connected nodes in the network. Importantly, EC serves as a measure of the connectivity against a fixed scale when normalized, so it can be used to reliably compare different networks [35]. For example, the normalization becomes essential when analyzing differences between graphs, for example, to study the pattern of centrality variation between the apo and holo states of a protein.

Katz centrality (KC) is a generalization of EC, which via two constants, namely an adjacency damping coefficient a and a basal adjacency b, assigns a centrality on the basis of a node's immediate connectivity. While b avoids adjacencies of zero, a weighs the magnitude of each centrality value. Node centrality can be dampened to various extents—larger values of a make KC tend towards EC.

$$KC_i = \alpha \sum_{j=1}^{n} A_{ij} KC_j + \beta \tag{12}$$

PageRank (PR) is an adjusted version of KC that also assigns node centrality based on that of their neighbors. For each round of the power iteration, the centrality of each neighbor to a node is normalized by its own degree (given that the graph is undirected), and each of the resulting neighbors' centrality is summed up and assigned to the parent node. As in *KC*, it also includes a damping factor α and a constant β.

$$PR_i = \alpha \sum_{j=1}^{n} \frac{A_{ij}}{D_j} PR_j + \beta \tag{13}$$

Eccentricity (*ECC*) is the longest path from a node to any other node in a graph.

$$\overline{ECC}_j = \frac{1}{m} \sum_{i=1}^{m} \max_{j,k \in V} d_i(j, k) \tag{14}$$

Like the PCN and the PSN, the AAN (Amino Acid Network) models the atomic structure of a protein by a network of amino acids in interactions, with the amino acids as nodes of the network and the atomic interactions as edges. Two amino acids i and j are in interaction or in contact when they have at least two atoms, one from i and one from j within a threshold Euclidian distance that covers chemical-interaction range; we use 5 Å but other thresholds can be used as well[45]. In the AAN, the link between two amino acids i and j is weighted by the exact number of atomic interactions between them. Both side chain and backbone atoms are considered. The difference with the PCN is the weight of the link and the fact that all atoms are considered, and the difference with the PSN is the absence of normalization of the link of the weight. The weight of a link (w_{ij}) characterizes the pair of amino acids in contact because it depends on the two amino acid features (side chain length, number atoms, volume, shape, etc.), their proximity and relative position as illustrated through simple schematics in Fig. 2a.

The node weight w_i is the sum of the node link weights (w_{ij}) over its k_i amino acid contacts. Compared to the node degree k_i, the node weight enables to assess the space occupied by atoms around an amino acid in the protein structure. Space can be occupied differently around amino acids having the same number of neighbors. For example, Fig. 2a shows amino acids with $k_i = 4$ around which space is occupied differently by their neighbors. A neighborhood is defined in terms of amino acid types, amino acid positions in the sequence and proximities thanks to the node weight.

We are interested in the dynamics and the tolerance to mutations and the AAN allows investigating that through the study of the space around amino acids (Fig. 2b). Packed neighborhoods (Fig. 2b, right schematic) leave little empty space available around an amino acid for atomic motions or/and for mutation tolerance compared to less packed neighborhoods (Fig. 2b, left and middle schematics). Computed over 736,149 amino acids from 750 protein structures, the average node weight (w_i/k_i), used to assess the average space occupied by atoms around amino acids, is moderate and of similar order ($<w_i/k_i>$ $\pm SD = 10 \pm 4$), regardless amino acid types and positions in the structure (Fig. 2c) [14]. This means that the local structure around amino acids is never too packed or too loose: surrounding atoms follow the Goldilocks principle, and empty space is always available for atomic motions and for mutation tolerance. We believe the geometrical constraint on the local space occupancy of moderate average link weight is one of the keys to the

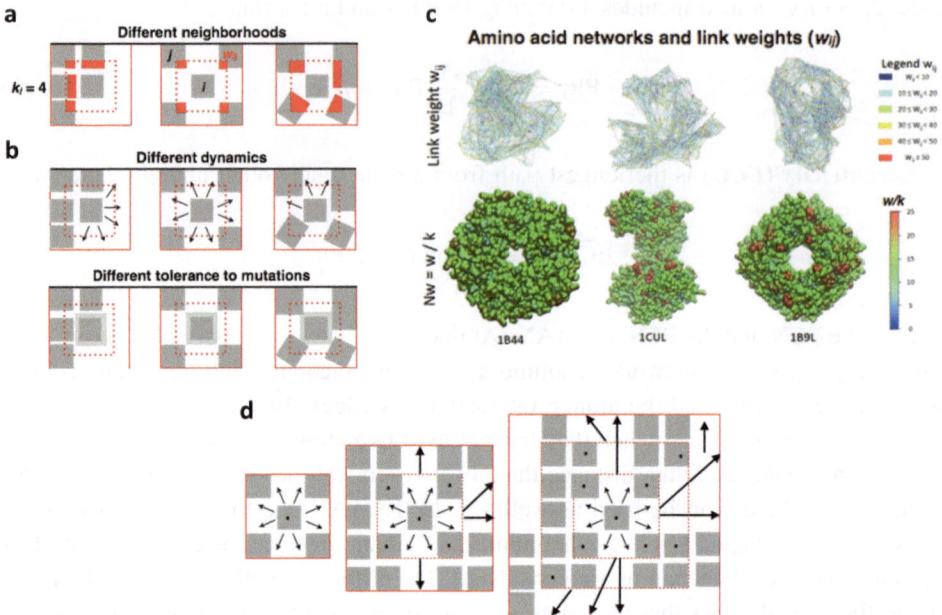

Fig. 2 Amino acid networks and the rational of link weights. **a** Schematics of the space occupancy by amino acids, represented by gray squares or rectangles, for nodes of degree 4 and different neighborhoods captured by link weights (w_{ij}, red area). The red dotted lines delineate the threshold distances within which atomic interactions (w_{ij}) are counted. 2D-spaces are shown for simplicity. **b** Schematics illustrating the interdependency between neighborhoods, dynamics and mutation tolerance. As in 2a, with arrows illustrating potential atomic motions within the free space available and mutation tolerance illustrated with the bigger light grey central square. **c** Three protein AANs (upper panel) and their respective PDB structures (lower panel, PDB code below) with amino acids in space-fill representation and colored according to their average node weight w/k values. **d** Schematics of space occupancy from the local scale to larger scales (red boxes) via different neighborhoods and potential multiple spatial scale dynamics embedded in the empty space remaining available (arrows)

design of a sustainable system that enables proteins to cope with mutations in time. Since the same geometrical constraint is built on an average value, there exist many alternative solutions to fulfill the constraint, implying a system sustainability that relies on diversity. Moreover, many alternatives to occupy the space around an amino acid introduce the possibility to carve the empty space locally and supervise the local dynamics. Then, across the AAN the local supervision outlines the different multi-scale dynamics underlying biological functions (Fig. 2d).

To test the hypothesis that the local space is carved to embed the multi-scale dynamics, we have monitored the weight and degree of the nodes of the AAN of the third PDZ domain of PDS-95 (PDB 1BE9) over multiple cuts-off and reported a large diversity of neighborhoods in terms of big, medium and small neighbors in agreement with the possibility of embedding different dynamics [38]. Moreover, we also observed that the node degree increases with the cut-off via only two types of behaviors: either a plateau or linearly. This indicates that large space occupancies are built again on common constraints (plateau/linear) produced from many alternative solutions as observed for the local space occupancy and the w/k values. The possibility to investigate some aspects of protein dynamics using the weight links of AAN is further illustrated by measuring changes upon mutation in the weight link allocation to the structural level of proteins, 1D, 2D, 3D and 4D which coincide with folding and unfolding differences associated with the mutations [7, 39]. Coupling AAN, weight links and MD is also efficient to analyze MD result and pinpoint residue motions key for allostery [1, 19, 20, 33]. More broadly, amino acid networks and the mathematical tools available to analyze them contribute to the understanding of protein structures and protein features (e.g. robustness to mutations) both in parallel and as a complement to experimental data [26] add new ref if possible [40]. Further exploration of network-based approaches and the tools they provide will help get a deeper understanding of protein performances such as smart motions or functional robustness and reveal how to design robust functional systems.

3   Molecular Descriptors Based on ENMs

In ENMs, each node represents a C_α atom in proteins and each edge is a spring γ for connecting two sites within a given cutoff distance r_c. The two most commonly used ENM methods [15, 28], the Gaussian network model (GNM) and the anisotropic network model (ANM) are introduced in this chapter. The total potential energy of the ANM and GNM systems with N nodes are expressed as:

$$V_{GNM} = -\frac{\gamma}{2}\left[\sum_{i=1}^{N-1}\sum_{j=i+1}^{N}\left(R_{ij}-R_{ij}^0\right)\cdot\left(R_{ij}-R_{ij}^0\right)\Gamma_{ij}\right] \tag{15}$$

$$V_{ANM} = -\frac{\gamma}{2}\left[\sum_{i=1}^{N-1}\sum_{j=i+1}^{N}\left(R_{ij}-R_{ij}^0\right)^2\Gamma_{ij}\right] \tag{16}$$

where R_{ij} and R_{ij}^0 are the instantaneous and equilibrium distances between nodes i and j, respectively, and Γ_{ij} is the ijth element of the $N \times N$ Kirchoff matrix Γ, which is written as:

$$\Gamma_{ij} = \begin{cases} -1 & i \neq j, R_{ij} \leq r_c \\ 0 & i \neq j, R_{ij} > r_c \\ -\sum_{i,i\neq j}\Gamma_{ij} & i=j \end{cases} \tag{17}$$

Normally, r_c between protein nodes were 7 Å and 13 Å for GNM and ANM, respectively. In comparison with the GNM, which only measures fluctuation, the ANM provides additional information on the motion directions of each residue.

The normal modes are extracted by eigenvalue decomposition $\Gamma = U \wedge U^T$. U is the orthogonal matrix whose kth column U_k is the kth mode eigenvector, and $\wedge$ is the diagonal matrix of eigenvalues, λ_k. **Mean-square fluctuations (*MSF*)** of a residue are given by

$$\langle (\Delta R_i)^2 \rangle = \frac{3k_B T}{\gamma} \sum_k \left[\left(U_k \wedge U_k^T \right)^{-1} \right]_{ii} \tag{18}$$

where k_B and T represent the Boltzmann constant and temperature, respectively. The other commonly used flexibility parameter in ENM is the so-called deformation energy [48]. The **deformation energy** for the ith residue, **$DE_i(k)$**, in the kth normal mode is defined as:

$$DE_i(k) = \sum_{j=1}^{n_{ci}} \frac{\gamma}{2} \left(\left| R_{ij}^0 + \Delta R_j(k) - \Delta R_i(k) \right| - \left| R_{ij}^0 \right| \right)^2 / N\lambda_k \tag{19}$$

where n_{ci} is the number of neighboring sites to the ith residue defined by the cutoff distance r_c, R_{ij}^0 is the equilibrium distance vector between residues i and j, $\Delta R_i(k)$ is the displacement vector of the ith mode at the kth normal mode and N is the total number of nodes in the elastic network. The deformation energy can also be calculated for a single mode or a weighted sum over several modes. Nodes with low deformation energy are corresponding to rigid residues, whereas flexible residues have high energy densities.

Applying perturbation response scanning theory to the ENM generates a novel **dynamic flexibility index (*DFI*)** for each residue [9].

$$DFI_j = \frac{\sum_{i=1}^{N} A_{ij}}{\sum_{j=1}^{N} \sum_{i=1}^{N} A_{ij}} \tag{20}$$

where A_{ij} is the element of the perturbation response matrix, which corresponds to the response fluctuation profile of residue j upon perturbation residues i.

Similar to the DFI, the **dynamic coupling index (*DCI*)** captures the strength of the displacement response of a given position i upon perturbation to a single functionally important position (or subset of positions) j, relative to the average fluctuation response of position i calculated using perturbations to all other positions within a structure:

$$DCI_i = \frac{\sum_j^{N_{functional}} \left| \Delta R^j \right|_i / N_{functional}}{\sum_{j=1}^{N} \left| \Delta R^j \right|_i / N} \tag{21}$$

As such, the DCI can be considered a measure of the dynamic coupling between residues i and j upon perturbation to residue j.

The Markov stochastic model coupled with ENM is used to explore the signal transduction of perturbations in proteins in terms of **hitting time** and **commute time**. In general, the hitting time for the transfer of a signal from residue j to i is given by:

$$H(i,j) = \sum_{k=1}^{N} \left\{ \left[\Gamma^{-1} \right]_{kj} - \left[\Gamma^{-1} \right]_{ij} - \left[\Gamma^{-1} \right]_{ki} - \left[\Gamma^{-1} \right]_{ii} \right\} \cdot d_k \qquad (22)$$

where Γ is the Kirchoff matrix obtained by GNM. The average hit time for the i-th residue $<H(i)>$ is the average of $H(i, j)$ over all starting points i. The commute time is defined by the sum of the hitting times in both directions, that is:

$$C(i,j) = H(i,j) + H(j,i) \qquad (23)$$

Accordingly, commute times provide a metric of the efficiency of allosteric communication.

Based on ENM calculations, several other kinds of matrices can be generated. From the perturbation response scanning (PRS) matrix [3], two dynamic descriptors of **effectiveness** and **sensitivity** were defined as row and column averages of the matrix, respectively. The effector residues most effectively propagate signals in response to external perturbations. The sensor residues can easily sense signals and respond with dynamic changes. Directly from the Kirchoff matrix, its eigenvalues can be used to ascertain how important each node is to maintain the overall mechanical connectedness of the network. This amounts to measuring how much the network Laplacian spectrum changes when the connections, or couplings, of a node with its neighbors are deleted. As such, the **mechanical bridging score (MBS)** for a given mutation reflects the response ability of this residue. In the symmetrical stiffness matrix, the elements describe the effective spring constants associated with each residue pair. The **stiffness** for individual mutational residues is obtained by averaging all the elements in the corresponding row/column of the matrix. These dynamic descriptors based on ENMs are summarized in Fig. 3, and a detailed description can be found in Bahar's work [42]. Their calculation can be performed by ProDy as a Python-based package [6], and DynOmics as an easily-used web server [27].

4　Applications in the Prediction of Functional Sites

In the AlphaFold2 era, the identification of functional sites is becoming the age, which is widely acknowledged as an important biological problem. In addition to normally used evolutionary descriptors, such as Shannon information entropy and mutual information, the above network-based molecular descriptors both at the structural and the dynamics levels provide more information for predicting protein functional sites. Here, we will

Fig. 3 Molecular descriptors based on ENMs. MSF: Mean-square fluctuation, DE: Deformation energy, MBS: Mechanical bridging score, DFI: Dynamic coupling index

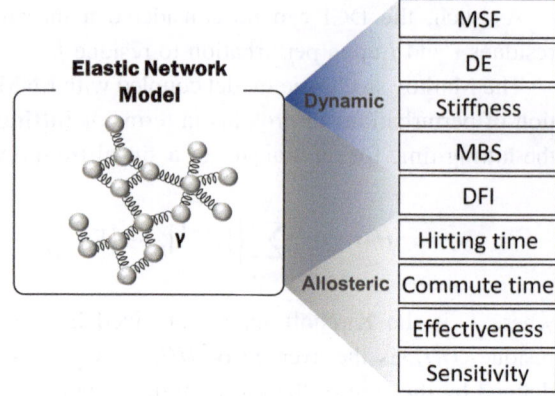

focus on the current applications in the development of machine learning models coupled with evolutionary descriptors for three types of functional sites: disease-related mutations, functional PTMs, and ligand-binding sites (Fig. 4a).

In 2020, Ponzoni et al. [43] first introduced a new machine learning method called Rhapsody that considers dynamic-based descriptors, including MSF, effectiveness, sensitivity, and stiffness to predict the pathogenicity of missense variants. The utility of

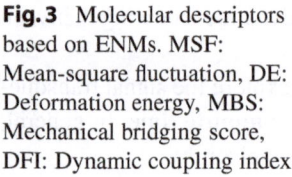

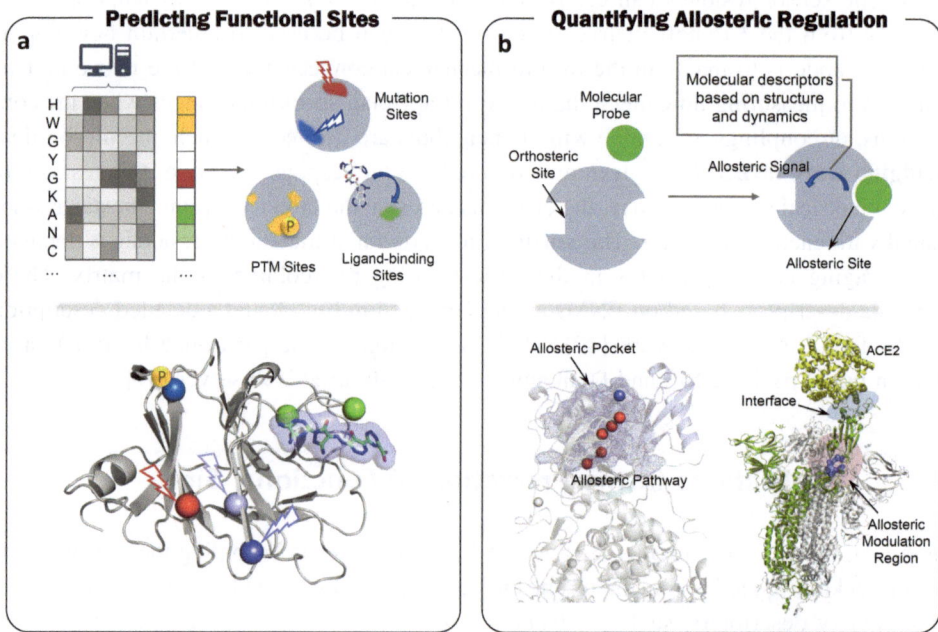

Fig. 4 Applications in predicting functional sites and quantifying allosteric regulation

Rhapsody and protein dynamics in predicting missense variants has been illustrated by the application to human H-Ras, phosphatase and tensin homolog and thiopurine S-methyltransferase, with their deep mutational scanning data. More recently, we have also applied a similar idea by incorporating network-based descriptors to investigate the mutational genotype-disease phenotype relationship [50]. In the case of tissue-nonspecific alkaline phosphatase (TNSALP), we first performed the statistical analysis between the molecular descriptors and TNSALP mutations with the control, mild and severe phenotypes. Then, the developed machine learning model suggested that the network descriptor of betweenness could serve as a robust indicator of severe mutations, showing the allosteric effects of the pathogenesis of mutations.

In another recent work, we also presented a framework by integrating the sequence, structural topology, and particular dynamics features to characterize the functional context and druggabilities of PTMs in the well-known kinase family [57]. The Machine learning models including network-based molecular descriptors could predict functional PTMs, whose potential role in drug design has also been investigated by covalent inhibitor targeting experiments. The prediction of ligand-binding sites is particularly useful for understanding drug design. By including a comprehensive set of structure-based network descriptors and four ENM-based dynamic descriptors including MSF, DFI, active site perturbation response, shortest dynamically correlated path to active site residues, Mishra et al. [34] developed discrete machine learning models (AR-Pred) using the random forest approach to predict allosteric and active site residues. The AR-Pred model yielded a median area under the curve (AUC) of 91% and a Matthews correlation coefficient (MCC) of 0.68 for ligand-binding site prediction, showing comparable performance to other existing methods.

5   Applications in Quantifying Allosteric Regulation

Network-based molecular descriptors provide a lot of mathematical tools to quantify allosteric regulation and measure their signal transduction [37]. Accordingly, we have proposed an integrated framework of using network-based descriptors for the high-throughput modeling of protein dynamics and allosteric regulation [30]. The method has been further applied to several protein systems successfully, and we will give two examples in the following (Fig. 4b).

DNA methyltransferase 1 (DNMT1), a large multidomain enzyme, and its versatile functions rely on allosteric networks between its different interacting partners. For the DNMT1 monomer, hinge sites predicted by MSF at the RFTS-CD interface are key regulators for inter-domain interactions; the study of effectiveness, sensitivity and shortest pathway based on PSN suggest both intra-domain allosteric networks and inter-domain communication pathways in DNMT1 [31]. In addition, based on the modeling structures of DNMT1-ubiquitylated H3 (H3Ub)/ubiquitin specific peptidase 7 (USP7) complexes, we also used a combination of structure- and dynamics- based molecular descriptors to examine their molecular mechanisms of allosteric regulation [59]. Network modeling and site perturbation- based descriptors, particularly network betweenness and effectiveness, have revealed the role of mutational hotspots and PTM sites in the allosteric interaction landscape in both DNMT1 protein–protein complexes.

The second example system is the SARS-spike glycoprotein-human angiotensin-converting enzyme 2 (S-ACE2) complex, which is critical to understand the molecular mechanisms underlying COVID-19 action. The comparison of the S-ACE2 of SARS-CoV-2 and SARS-CoV suggests an allosteric modulation region in the SARS-CoV-2 spike protein [12]. By using two network parameters based on the spectral clustering of PCNs, intramodule connectivity z_i and participation coefficient P_i, we suggested a putative allosteric site in the SARS-CoV-2 spike protein. Then, the allosteric ability of this site was also highlighted by the effectiveness based on ENM modeling. We have further applied PCN and ENM to quantify the allosteric regulation of free (closed and open) states of spike protein and spike–ACE2 complexes [22]. The effectiveness calculation showed that the spike protein in the ACE2-bound state shows higher allosteric potential than that in the closed and open states, and the MSF calculation found a correspondence between hinge sites and the AMR in the S-ACE complex, suggesting a molecular basis for hepcidin involvement in COVID-19 pathogenesis.

6 Conclusions

Molecular descriptors, as a core aim in mathematical chemistry [18], also provide a rich toolbox to quantify and understand complex behavior in proteins underlying their dynamics and allosteric regulation. In addition, the sheer scope and completeness of predicted protein structures in the AlphaFold database (https://alphafold.ebi.ac.uk) offers an alternative form of value, particularly favoring functional studies at the proteome level. Here, we have outlined some network-based molecular descriptors both at the structural and dynamics levels, along with their bioinformatics tools. The utilization of these molecular descriptors in decrypting protein dynamics highlights the practical applications of functional site prediction and allosteric regulation quantification. Despite this progress, we think that compared with molecular descriptors in mathematical chemistry, network-based molecular descriptors for identifying "central" interactions, modularity, and dynamics are still needed.

Acknowledgements This research was funded by the National Natural Science Foundation of China (32271292, 31872723), the Priority Academic Program Development (PAPD) of Jiangsu Higher Education Institutions.

References

1. Aledavood E, Gheeraert A, Forte A, Vuillon L, Rivalta I, Luque FJ, Estarellas C (2021) Elucidating the activation mechanism of AMPK by direct pan-activator PF-739. Front Mol Biosci 8:760026. https://doi.org/10.3389/fmolb.2021.760026

2. Ashtiani M, Salehzadeh-Yazdi A, Razaghi-Moghadam Z, Hennig H, Wolkenhauer O, Mirzaie M, Jafari M (2018) A systematic survey of centrality measures for protein-protein interaction networks. BMC Syst Biol 12(1):80. https://doi.org/10.1186/s12918-018-0598-2

3. Atilgan C, Atilgan AR (2009) Perturbation-response scanning reveals ligand entry-exit mechanisms of ferric binding protein. Plos Comput Biol 5(10):e1000544. https://doi.org/10.1371/journal.pcbi.1000544

4. Baek M, DiMaio F, Anishchenko I, Dauparas J, Ovchinnikov S, Lee GR, Wang J, Cong Q, Kinch LN, Schaeffer RD, Millan C, Park H, Adams C, Glassman CR, DeGiovanni A, Pereira JH, Rodrigues AV, van Dijk AA, Ebrecht AC, Opperman DJ, Sagmeister T, Buhlheller C, Pavkov-Keller T, Rathinaswamy MK, Dalwadi U, Yip CK, Burke JE, Garcia KC, Grishin NV, Adams PD, Read RJ, Baker D (2021) Accurate prediction of protein structures and interactions using a three-track neural network. Science 373(6557):871–+. https://doi.org/10.1126/science.abj8754

5. Bagdonas H, Fogarty CA, Fadda E, Agirre J (2021) The case for post-predictional modifications in the AlphaFold protein structure database. Nat Struct Mol Biol 28(11):869–870. https://doi.org/10.1038/s41594-021-00680-9

6. Bakan A, Meireles LM, Bahar I (2011) ProDy: protein dynamics inferred from theory and experiments. Bioinformatics 27(11):1575–1577. https://doi.org/10.1093/bioinformatics/btr168

7. Bourgeat L, Pacini L, Serghei A, Lesieur C (2021) Experimental diagnostic of sequence-variant dynamic perturbations revealed by broadband dielectric spectroscopy. Structure 29(12):1419–1429 e1413. https://doi.org/10.1016/j.str.2021.05.005

8. Brown DK, Penkler DL, Sheik Amamuddy O, Ross C, Atilgan AR, Atilgan C, Tastan Bishop O (2017) MD-TASK: a software suite for analyzing molecular dynamics trajectories. Bioinformatics 33(17):2768–2771. https://doi.org/10.1093/bioinformatics/btx349

9. Campitelli P, Modi T, Kumar S, Ozkan SB (2020) The role of conformational dynamics and allostery in modulating protein evolution. Annu Rev Biophys 49:267–288. https://doi.org/10.1146/annurev-biophys-052118-115517

10. Di Paola L, De Ruvo M, Paci P, Santoni D, Giuliani A (2013) Protein contact networks: an emerging paradigm in chemistry. Chem Rev 113(3):1598–1613. https://doi.org/10.1021/cr3002356

11. Di Paola L, Giuliani A (2015) Protein contact network topology: a natural language for allostery. Curr Opin Struct Biol 31:43–48. https://doi.org/10.1016/j.sbi.2015.03.001

12. Di Paola L, Hadi-Alijanvand H, Song X, Hu G, Giuliani A (2020) The discovery of a putative allosteric site in the SARS-CoV-2 spike protein using an integrated structural/dynamic approach. J Proteome Res 19(11):4576–4586. https://doi.org/10.1021/acs.jproteome.0c00273

13. Doncheva NT, Klein K, Domingues FS, Albrecht M (2011) Analyzing and visualizing residue networks of protein structures. Trends Biochem Sci 36(4):179–182. https://doi.org/10.1016/j.tibs.2011.01.002

14. Dorantes-Gilardi R, Bourgeat L, Pacini L, Vuillon L, Lesieur C (2018) In proteins, the structural responses of a position to mutation rely on the Goldilocks principle: not too many links, not too few. Phys Chem Chem Phys 20(39):25399–25410. https://doi.org/10.1039/c8cp04530e

15. Eyal E, Lum G, Bahar I (2015) The anisotropic network model web server at 2015 (ANM 2.0). Bioinformatics 31(9):1487–1489. https://doi.org/10.1093/bioinformatics/btu847

16. Fauser J, Leschinsky N, Szynal BN, Karginov AV (2022) Engineered allosteric regulation of protein function. J Mol Biol 434(17):167620. https://doi.org/10.1016/j.jmb.2022.167620
17. Felline A, Seeber M, Fanelli F (2022) PSNtools for standalone and web-based structure network analyses of conformational ensembles. Comput Struct Biotechnol J 20:640–649. https://doi.org/10.1016/j.csbj.2021.12.044
18. Fernandez-Torras A, Comajuncosa-Creus A, Duran-Frigola M, Aloy P (2022) Connecting chemistry and biology through molecular descriptors. Curr Opin Chem Biol 66:102090. https://doi.org/10.1016/j.cbpa.2021.09.001
19. Gheeraert A, Pacini L, Batista VS, Vuillon L, Lesieur C, Rivalta I (2019) Exploring allosteric pathways of a V-type enzyme with dynamical perturbation networks. J Phys Chem B 123(16):3452–3461. https://doi.org/10.1021/acs.jpcb.9b01294
20. Gheeraert A, Vuillon L, Chaloin L, Moncorge O, Very T, Perez S, Leroux V, Chauvot de Beauchene I, Mias-Lucquin D, Devignes MD, Rivalta I, Maigret B (2022) Singular interface dynamics of the SARS-CoV-2 delta variant explained with contact perturbation analysis. J Chem Inf Model 62(12):3107–3122. https://doi.org/10.1021/acs.jcim.2c00350
21. Guzzi PH, di Paola L, Giuliani A, Veltri P (2022) PCN-Miner: an open-source extensible tool for the analysis of protein contact networks. Bioinformatics. https://doi.org/10.1093/bioinformatics/btac450
22. Hadi-Alijanvand H, Di Paola L, Hu G, Leitner DM, Verkhivker GM, Sun PX, Poudel H, Giuliani A (2022) Biophysical Insight into the SARS-CoV2 Spike-ACE2 interaction and its modulation by hepcidin through a multifaceted computational approach. ACS Omega 7(20):17024–17042. https://doi.org/10.1021/acsomega.2c00154
23. Halder A, Anto A, Subramanyan V, Bhattacharyya M, Vishveshwara S, Vishveshwara S (2020) Surveying the Side-chain network approach to protein structure and dynamics: the SARS-CoV-2 spike protein as an illustrative case. Front Mol Biosci 7:596945. https://doi.org/10.3389/fmolb.2020.596945
24. Hu G (2021) Identification of allosteric effects in proteins by elastic network models. Methods Mol Biol 2253:21–35. https://doi.org/10.1007/978-1-0716-1154-8_3
25. Jumper J, Evans R, Pritzel A, Green T, Figurnov M, Ronneberger O, Tunyasuvunakool K, Bates R, Zidek A, Potapenko A, Bridgland A, Meyer C, Kohl SAA, Ballard AJ, Cowie A, Romera-Paredes B, Nikolov S, Jain R, Adler J, Back T, Petersen S, Reiman D, Clancy E, Zielinski M, Steinegger M, Pacholska M, Berghammer T, Bodenstein S, Silver D, Vinyals O, Senior AW, Kavukcuoglu K, Kohli P, Hassabis D (2021) Highly accurate protein structure prediction with AlphaFold. Nature 596(7873):583–589. https://doi.org/10.1038/s41586-021-03819-2
26. Lesieur C, Vuillon L (2021) Topology results on adjacent amino acid networks of oligomeric proteins. Methods Mol Biol 2253:113–135. https://doi.org/10.1007/978-1-0716-1154-8_8
27. Li H, Chang YY, Lee JY, Bahar I, Yang LW (2017) DynOmics: dynamics of structural proteome and beyond. Nucleic Acids Res 45(W1):W374–W380. https://doi.org/10.1093/nar/gkx385
28. Li H, Chang YY, Yang LW, Bahar I (2016) iGNM 2.0: the Gaussian network model database for biomolecular structural dynamics. Nucleic Acids Res 44(D1):D415–422. https://doi.org/10.1093/nar/gkv1236
29. Li H, Doruker P, Hu G, Bahar I (2020) Modulation of toroidal proteins dynamics in favor of functional mechanisms upon ligand binding. Biophys J 118(7):1782–1794. https://doi.org/10.1016/j.bpj.2020.01.046
30. Liang ZJ, Verkhivker GM, Hu G (2020) Integration of network models and evolutionary analysis into high-throughput modeling of protein dynamics and allosteric regulation: theory, tools and applications. Brief Bioinform 21(3):815–835. https://doi.org/10.1093/bib/bbz029

31. Liang ZJ, Zhu Y, Long J, Ye F, Hu G (2020) Both intra and inter-domain interactions define the intrinsic dynamics and allosteric mechanism in DNMT1s. Comput Struct Biotec 18:749–764. https://doi.org/10.1016/j.csbj.2020.03.016

32. Liu C, Ma YF, Zhao J, Nussinov R, Zhang YC, Cheng FX, Zhang ZK (2020) Computational network biology: data, models, and applications. Phys Rep 846:1–66. https://doi.org/10.1016/j.physrep.2019.12.004

33. Maschietto F, Gheeraert A, Piazzi A, Batista VS, Rivalta I (2022) Distinct allosteric pathways in imidazole glycerol phosphate synthase from yeast and bacteria. Biophys J 121(1):119–130. https://doi.org/10.1016/j.bpj.2021.11.2888

34. Mishra SK, Kandoi G, Jernigan RL (2019) Coupling dynamics and evolutionary information with structure to identify protein regulatory and functional binding sites. Proteins 87(10):850–868. https://doi.org/10.1002/prot.25749

35. Negre CFA, Morzan UN, Hendrickson HP, Pal R, Lisi GP, Loria JP, Rivalta I, Ho J, Batista VS (2018) Eigenvector centrality for characterization of protein allosteric pathways. Proc Natl Acad Sci U S A 115(52):E12201–E12208. https://doi.org/10.1073/pnas.1810452115

36. Ni D, Liu Y, Kong R, Yu Z, Lu S, Zhang J (2022) Computational elucidation of allosteric communication in proteins for allosteric drug design. Drug Discovery Today 27(8):2226–2234. https://doi.org/10.1016/j.drudis.2022.03.012

37. Nussinov R, Tsai CJ, Jang H (2022) Allostery, and how to define and measure signal transduction. Biophys Chem 283.https://doi.org/10.1016/J.Bpc.2022.106766

38. Pacini L, Dorantes-Gilardi R, Vuillon L, Lesieur C (2021) Mapping function from dynamics: future challenges for network-based models of protein structures. Front Mol Biosci 8:744646. https://doi.org/10.3389/fmolb.2021.744646

39. Pacini L, Lesieur C (2021) A computational methodology to diagnose sequence-variant dynamic perturbations by comparing atomic protein structures. Bioinformatics. https://doi.org/10.1093/bioinformatics/btab736

40. Pacini L, Lesieur C (2022) GCAT: a network model of mutational influences between amino acid positions in PSD95(pdz3). Front Mol Biosci 9:1035248. https://doi.org/10.3389/fmolb.2022.1035248

41. Penkler DL, Atilgan C, Tastan Bishop O (2018) Allosteric modulation of human Hsp90alpha conformational dynamics. J Chem Inf Model 58(2):383–404. https://doi.org/10.1021/acs.jcim.7b00630

42. Ponzoni L, Bahar I (2018) Structural dynamics is a determinant of the functional significance of missense variants. P Natl Acad Sci USA 115(16):4164–4169. https://doi.org/10.1073/pnas.1715896115

43. Ponzoni L, Penaherrera DA, Oltvai ZN, Bahar I (2020) Rhapsody: predicting the pathogenicity of human missense variants. Bioinformatics 36(10):3084–3092. https://doi.org/10.1093/bioinformatics/btaa127

44. Rauer C, Sen N, Waman VP, Abbasian M, Orengo CA (2021) Computational approaches to predict protein functional families and functional sites. Curr Opin Struct Biol 70:108–122. https://doi.org/10.1016/j.sbi.2021.05.012

45. Salamanca Viloria J, Allega MF, Lambrughi M, Papaleo E (2017) An optimal distance cutoff for contact-based protein structure networks using side-chain centers of mass. Sci Rep 7(1):2838. https://doi.org/10.1038/s41598-017-01498-6

46. Sheik Amamuddy O, Afriyie Boateng R, Barozi V, Wavinya Nyamai D, Tastan Bishop O (2021) Novel dynamic residue network analysis approaches to study allosteric modulation: SARS-CoV-2 M(pro) and its evolutionary mutations as a case study. Comput Struct Biotechnol J 19:6431–6455. https://doi.org/10.1016/j.csbj.2021.11.016

47. Sheik Amamuddy O, Glenister M, Tshabalala T, Tastan Bishop O (2021) MDM-TASK-web: MD-TASK and MODE-TASK web server for analyzing protein dynamics. Comput Struct Biotechnol J 19:5059–5071. https://doi.org/10.1016/j.csbj.2021.08.043

48. Uyar A, Kurkcuoglu O, Nilsson L, Doruker P (2011) The elastic network model reveals a consistent picture on intrinsic functional dynamics of type II restriction endonucleases. Phys Biol 8(5):056001. https://doi.org/10.1088/1478-3975/8/5/056001

49. Verkhivker GM, Agajanian S, Hu G, Tao P (2020) Allosteric regulation at the crossroads of new technologies: multiscale modeling, networks, and machine learning. Front Mol Biosci 7. https://doi.org/10.3389/Fmolb.2020.00136

50. Xiao F, Zhou Z, Song X, Gan M, Long J, Verkhivker G, Hu G (2022) Dissecting mutational allosteric effects in alkaline phosphatases associated with different Hypophosphatasia phenotypes: an integrative computational investigation. Plos Comput Biol 18(3):e1010009. https://doi.org/10.1371/journal.pcbi.1010009

51. Yan W, Hu G, Liang Z, Zhou J, Yang Y, Chen J, Shen B (2018) Node-weighted amino acid network strategy for characterization and identification of protein functional residues. J Chem Inf Model 58(9):2024–2032. https://doi.org/10.1021/acs.jcim.8b00146

52. Yan W, Yu C, Chen J, Zhou J, Shen B (2020) ANCA: a web server for amino acid networks construction and analysis. Front Mol Biosci 7:582702. https://doi.org/10.3389/fmolb.2020.582702

53. Yan W, Zhang D, Shen C, Liang Z, Hu G (2018) Recent advances on the network models in target-based drug discovery. Curr Top Med Chem 18(13):1031–1043. https://doi.org/10.2174/1568026618666180719152258

54. Yan W, Zhou J, Sun M, Chen J, Hu G, Shen B (2014) The construction of an amino acid network for understanding protein structure and function. Amino Acids 46(6):1419–1439. https://doi.org/10.1007/s00726-014-1710-6

55. Yan WY, Hu G (2022) Structural biology meets biomolecular networks: the post-AlphaFold era. Curr Bioinform 17(6):493–497. https://doi.org/10.2174/1574893617666220211115211

56. Zha J, Li M, Kong R, Lu S, Zhang J (2022) Explaining and predicting allostery with allosteric database and modern analytical techniques. J Mol Biol 434(17):167481. https://doi.org/10.1016/j.jmb.2022.167481

57. Zhang H, He J, Hu G, Zhu F, Jiang H, Gao J, Zhou H, Lin H, Wang Y, Chen K, Meng F, Hao M, Zhao K, Luo C, Liang Z (2021) Dynamics of post-translational modification inspires drug design in the kinase family. J Med Chem 64(20):15111–15125. https://doi.org/10.1021/acs.jmedchem.1c01076

58. Zhang Y, Doruker P, Kaynak B, Zhang S, Krieger J, Li HC, Bahar I (2020) Intrinsic dynamics is evolutionarily optimized to enable allosteric behavior. Curr Opin Struct Biol 62:14–21. https://doi.org/10.1016/j.sbi.2019.11.002

59. Zhu Y, Ye F, Zhou ZY, Liu WL, Liang ZJ, Hu G (2021) Insights into conformational dynamics and allostery in DNMT1-H3Ub/USP7 interactions. Molecules 26(17). https://doi.org/10.3390/Molecules26175153